Table

M000103271

Table of Figures

Dedication

The Great Unconformity is dedicated
to the rise of activists within the millennial generation.
May you be the sustainable future of our planet.
With this book I send you merrily on your way
but only with your eyes open wide.

Acknowledgements

Quite simply, this book would not have been possible without my life partner, Bill Hanson. He has edited and sweated right along with me on each major re-write. Once Bill even dedicated two weeks of non-stop editing as I cranked through the second re-write. Working so intensely side by side reminded me of being back in college when you had to pull the all-nighters to get in all five end-of-the-year papers. You can only do it if your roommate is there with you. Similarly, you can only write a book like this if your lifetime roommate is with you…every step of the way. Thanks, Hon.

The third and final re-write got cheered and edited along by Rose Alexandre-Leach with Green Writer's Press. She encouraged and challenged me with new chapters and provided sound edits. Much appreciated.

Taking the manuscript to the end is my friend, producer and publisher Manuel "Max" Freedman. Max's belief in me as a person and as a writer of screenplays certainly spilled over into getting *The Great Unconformity* out into the world. Next, be watching for the movie, *Hot Ice, Cold Lies*.

Then there are those people that came along at just the right moment to nudge me along. The first was Katrina Woolford, a local publisher, who told me over lunch, "Yes, for sure, there's a book in your essays." Next, came fellow writers who encouraged me to evolve the book into more of a memoir. Then they turned around and gave me constant support throughout the process of getting published. Awesome to be part of Alaska's 49 Writers community.

The first photo that came my way for this book was "Two Whales Breaching". My friend Cameron Byrnes posted it on Facebook. It came just as I finished touching up "Whales, Waves and Wonder". What a sign of encouragement this was.

I am particularly grateful for award-winning photographer Mark Kelley who graciously let me use his stunning aurora photos. The cover photo also has some synchronistic elements to it. Although the auroras were not as bright when I looked out from my living room window and wrote the poem in the last chapter, it was indeed the same night this photo was taken and right near where I live. Perfect fit for the cover.

I started this acknowledgement with family and it ends with family. What a treat to have my brothers, Tim and Ray Troll, both contribute their art work. Their art adorns my home, why not my book? My brothers are awesome.

Last is my dog Nellie who sat at my feet patiently waiting for her walk while I typed.

—*Kate Troll, Douglas, Alaska, Spring 2017*

FOREWORD

THE GREAT UNCONFORMITY:
Reflections on Hope in an Imperiled World

Overall, the last few hundred years have wrought an extraordinary physical disengagement of humans from the natural environment and an increase of lack of civility among people. This shift is especially pronounced for people in westernized and/or industrialized countries—those that have the greatest historical culpability for environmental destruction. These westernized cultures exhibit symptoms of attachment disorder with nature. This attachment disorder may relate to common Western child rearing practices, including a lack of time in nature (D'Amore & Mitten, 2015). People have spent over 99% of their evolutionary history in close proximity to if not in nature. Though the proximity has changed, there is still an attachment with nature; but now it is a disordered attachment. This attachment disorder is exemplified in humans' violent relationships with the natural world; when we exhibit violence towards anything we exhibit violence towards ourselves.

Media exacerbates this estrangement from nature by relentlessly reporting about horrors that nature supposedly imposes upon us. Headlines such as "The 15 most horrible ways mother nature kills" or "The U.S. death map: Where and how nature kills most" give the sense that nature is purposefully hurting people. Even phrases such as 'fight climate change' encourage conflictual relationships. Combining media reports with the fact that many people in developed countries spend little time in direct contact with nature helps people become nature-phobic. Exceptions to isolation from nature occur in the midst of natural disasters: earthquakes, fires, tornadoes, cyclones, hurricanes, flooding, mudslides, and tidal waves. Considering the impact such encounters imprint on those who experience them, it is not surprising that many people react with fear and mistrust of the natural world. People witness and experience what they see and feel as violence towards them from the natural world.

But Nature is merely reacting to human actions. Through the severe weather happenings due to climate change, it seems that our relationship with the natural world is coming full circle from a time when we lived daily and intimately in the outdoors to now having the natural environment pressing around us, literally knocking at our doors, seemingly asking us to pay attention. Perhaps it is time for humans to play nicely. We are entangled with nature and the study of ecology is a study about relationships.

Relationships are central to every dimension of our lives and shape our lives because we are social, relational beings. Feedback loops continually operate in relationships, meaning that the more positive encounters we have with nature the more we come to expect them to be positive. The same is true for negative encounters. Our understanding of relationships and skill in developing and maintaining relationships is central to sustainability.

Borrowing from research about human relationships we can extend these concepts to humans' relationships with the natural environment. Gottman (2010) gained notoriety studying what causes relationships to fail, namely contempt, criticism, and defensiveness. We see evidence of these in humans' reactions to and interactions with nature. Thinking that one is more knowledgeable or more intelligent than others or making statements to another from a superior place is a form of contempt. Examples of contemptuous and critical writing about nature abound.

In order to engage in healthy relationships, Gottman identified the importance of maintaining communication that includes appreciative, reciprocal, and compassionate exchanges as a way to build positive perspective about another person—or about the natural world. If people see the natural world as more positive than negative, then when events happen that might be interpreted as negative they are more likely to be repaired or overlooked, just as in human relationships. An example of repair with the natural world might be seeing a rainbow and feeling content or hope. Many cultures see this harmonic combination of light as symbolizing beauty, promise, and enlightenment.

Discovering how to be in healthy relationships is key to the overall health and sustainability of the planet and beyond and helps guide interactions with each other and the environment. We can learn from indigenous people who have embedded traditional wisdom and practices that honor the reciprocity between people and their natural surroundings. Perhaps all humans have an internal or intra-indigenous consciousness or psyche that can be awakened or activated through time in the outdoors and direct experiences with nature. Outdoor experiences can help change a narrative of fear and competition to one of mutualism and belonging, which influence our ecological identity and environmental behaviors. We may feel on a visceral level our spiritual and emotional connections with nature and regain our healthy attachment.

Kate has traveled and journeyed in many different outdoor environments and does so through an appreciative lens. In deep relationship with the natural environment and being self- and mutually-reliant, she models how to maintain healthy emotional and spiritual connections with nature, and that fuels her political and organizing work. She maintains her commitment to work hard for pro-environmental issues perhaps because of her sense of place and her close physical and spiritual connection to nature. Kate sustains her positive relationship with nature by playing in and with nature.

In *The Great Unconformity*, Kate blends hope and reality. Kate's adventures and her environmental work reflect how our lives are intertwined with all beings and natural systems as an ecology of relationships, which Ellen Swallow Richards understood in the late nineteenth century when she named the discipline of ecology (the first term she used was oekologie and shortly after that she used ecology). Swallow Richards saw humans in relationship with all things in and on Earth and sought to educate people that we need to change the way we relate to the environment in order to maintain the health of the environment—and our own health.

Kate demonstrates the inquisitiveness described by Rachel Carson's article "The Sense of Wonder" (2011) in which she encouraged humans to wonder more and to approach nature and life with curiosity and a sense of awe. In order to engage in healthy relationships with nature, we need to be, like Kate, healthily attached. Kate's bond with the land seems to have a sense of reverence that gives her an intrinsic motivation to treat nature with respect, and through an environmental ethic of care has an unwavering commitment to help people develop practical solutions for environmental concerns, which in turn has greatly influenced Alaska's political landscape, and landscapes elsewhere, in regards to the environment.

It is an honor to write a foreword to this book. Having run an outdoor tripping organization focused on the environment and women, I believe that I know about following convictions and dreams in one's work endeavors. There were no books or writings about "how to do" this sort of education, but that was a good thing. Our personal philosophies about leadership were relational, and we led with values of caring, respect, and a delight in beauty and creative expression. We used our experiences and our imaginations. Through being in a supportive group atmosphere, and for some re-defining their relationship with nature, I saw countless women gain a sense of renewal, relax, feel empowered, have fun, network, and find spiritual healing. It can be a time of learning and improving one's outdoor and living skills, of adventure and making new friends, and also can be healing, help to provide insight regarding one's interpersonal relationships, or provide clarity for decisions or transitions one may be faced with. Kate lives this relationship with nature.

Laws of science recognize that all human activity is inherently entangled in the natural world, which sustains us. Kate shows through her work and play how her healthy attachment influences her decisions to work with diverse groups of people to create win-win solutions. We can follow her lead. Much has happened in the last 30 years to deprive us of hope and to maximize concern. However, Kate's example of dedicated, relentless work while she keeps playing in nature can serve as a gentle reminder of our spiritual connection to nature and our ability to make a difference.

Denise Mitten, Ph.D.

Co-founder of Woodswomen, Inc. and Chair of the Sustainability Education Doctoral Program at Prescott College

Carson, R. (2011). *The Sense of Wonder* [Ebook]. Retrieved from http://www. openroadmedia.com

D'Amore, C. & Mitten, D. (2015). Nurtured Nature: The Connection Between Care for Children and Care for the Environment, in Thomas, P.L., Carr, P., Gorlewski, J., & Porfilio, B.(Eds) *Pedagogies of Kindness and Respect: On the Lives and Education of Children.* New York, NY: Peter Lang.

Gottman, J. (2010, January 5). Making Relationships Work. Retrieved from https:// www.youtube.com/watch?v=9aSpl_ZjmcY&spfreload=10

PART ONE   INTRODUCTION

I live in one of the best places, bar none, to appreciate the wild natural environment. I also live in one of the most politically difficult places to work on its behalf: Alaska. I arrived in Alaska thirty-nine years ago with a Master's degree in Natural Resource Management from the Yale School of Forestry and Environmental Studies. I arrived wanting to learn from the land management and environmental mistakes of the Lower Forty-eight states and do things right in this powerful landscape that I would make my home. But Alaska had a few things to teach me first. From my initial professional trial by fire trying to reconcile logging with a world-class eagle habitat in a small community, to helping convince former Governor Sarah Palin to promote renewable energy, the Alaska School of Hard Knocks has been my calling.

I live with a personal sense of wonder in the presence of mountains, rivers, and oceans. It is only by immersing myself headlong into wilderness that I know I belong to something far vaster than my career. I've learned that touching this wonder sustains my ability to work on behalf of the environment no matter how difficult the politics may be.

For the last decade, climate change has been my primary focus as Alaska has been heating up at more than twice the rate of the continental U.S. One reason for this is the *albedo effect*. Albedo is the fraction of solar energy that is reflected from the Earth back into space. It is a measure of the reflectivity of the earth's surface. Snow and ice reflect 85 percent to 90 per cent of the sun's energy, while land and open water absorb much of this energy. When the earth warms, the highly reflective snow and ice are replaced by darker land, vegetation, and open water, which further increases the absorption of heat in an unending feedback loop. Because of the albedo effect, the glaciers are in rapid retreat, polar sea ice is quickly shrinking, and permafrost (permanently frozen soil typically found a few feet below the surface) is melting. From the Arctic to Southeast Alaska, where I live, the physical and biological changes associated with climate change have been profound.

If any place should have a climate action plan, it is Alaska. From Mexico to British Columbia, every state and Canadian provincial government along the Pacific coast is engaged in climate action. But Alaska is the one region left behind.

This is the dichotomy of Alaska from which I write the political threads of this book. While my political experience is rooted in Alaska, I've worked nationally and internationally to bring environmental and economic considerations together in solutions for communities and industries. I know all too well how the issues of sustainability and climate change are global in nature. In *The Great Unconformity* I bring the inspiration and insight of global thinkers—Al Gore, Gus Speth and Naomi Klein to name a few—into my view of the world. I am not shy about relying on the great minds of today, nor the nature philosophers of yesteryear: Emerson, Muir, and Leopold. For it is only through collective wisdom and spiritual uplift that today's advocates will be able to sustain the effort necessary to begin healing the Earth.

The Great Unconformity is equal parts political analysis and personal reflection. The threads of this memoir build upon my contributions to conservation successes and defeats. I have found that the synergy successes between development and environmental conservation rise and fall like the tide in the saltwater channel outside my window. From the ebb and flood of experiences in an unconventional career, what are the lessons to help future generations? From the top of Denali to the bottom of the Grand Canyon, what is it about my sideways journeys and adventures, both internal and external that lend useful insight?

Woven throughout my stories are the currents of worry and wonder that tug at each other. How can we create sustainable communities as the world's population balloons over seven billion people, all striving for a middle class lifestyle? What happens if climate change continues unabated? How can we begin to heal our Earth from the mistakes of the past and the mistakes yet to come? Is the 2015 Paris climate agreement enough? Will the clean energy revolution come soon enough? For me, the only medicine strong enough to counter these worries is the wonder of living in nature, a medicine that Alaska doses out like no other place.

Along with the weight of the world and the wonder of nature, belief in the rise of the global mind weaves throughout this book. So do "hope spots." The notion of "hope spots" comes from renowned marine conservationist, Dr. Sylva Earle, who identified key ocean habitats that are critical to the future health of our world's oceans.[1] I've expanded this notion of hope spots to include policy areas where change for the better is occurring.

In a red state like Alaska, the hope spots can be hard to find but they are there buried in the avalanche of oil and gas politics. Here on the frontline of rapid change, it's easy to get overwhelmed and discouraged wherein each hope spot is akin to finding a nugget of gold. It is much easier to look elsewhere for progress on the climate front, to look to green states, to Europe and beyond. So this memoir moves beyond Alaska and into the realm of promising technological and policy developments. It also touches upon important cultural shifts and changes in attitudes such as the growing acceptance and role of women leaders.

In an article, "Ten Ideas for Saving the Planet", John B. Cobb, a Christian ethnologist notes, "Those of us who have had good fortune to encounter a better way of understanding the world have a profound responsibility to share it."[2] Alaska has been my lens and my motivation for finding a better way of understanding the world. You can judge whether or not I found it but nonetheless I write this book to fulfill my responsibility for sharing my good fortune... my good fortune to live in Alaska while encountering wisdom about sustainability and climate change.

Mr. Cobb developed an eco-theology based on the environmental philosophy of Alfred North Whitehead. Mr. Cobb's theology is based on the fundamental idea "that, despite trends to the contrary, we might avoid destroying the life-support systems on which we and other living beings depend." From this foundation Mr. Cobb elaborates ten interlocking ideas. His first idea is that "reality is composed of interrelated events," and his tenth idea is that "every community should be part of a community of communities." He concludes "the most important form of power is that which empowers others. A world in which that is deeply understood can be a sustainable world."

I understand deeply this connection and hope this book empowers you. From sharing policy insights and analysis there is a lot of material to digest. However, by wrapping the current events of sustainability and climate change in a cloak of stories about my personal and professional adventures, it is my hope that even the casual activist follows along, enjoys the ride and leaves inspired. Here is the first of many stories to come.

Chapter 1 — Whales, Worry, and Wonder

Figure 1: The wonderful and rarely ever seen event of two whales breaching at the same time. Photo Courtesy of Cameron Byrnes, Juneau, AK, 2011.

On New Year's Day, 2006, I walked with my husband, Bill, to one of our favorite places in Juneau, a peninsula surrounded by beaches on three sides that juts out into Lynn Canal, a fjord not far from our home. It was a glorious day of broken clouds and sunshine. As we emerged out of the woods, I gazed at the "godlight"—rays of sunlight fractured by the brooding clouds, streaming down over the snow-capped mountains.

Moved by this sight, I mused to Bill, "Wow. Maybe this is a sign. Maybe this will be the year when the truth about the Iraq war gets more media than all the fear-mongering rhetoric. Maybe this will be the year when ethics matter again." Now on a roll, I saluted the grandeur and wished, "Maybe this will be the year when the political pendulum finally starts to swing back and issues like health care, helping the poor and the environment might get attention."

"Well, it won't happen on the environment," replied Bill, always one to inject reality and bring me back to earth.

"But even Senator McCain gets global warming and has introduced a bill on climate change," I responded.

"With all the rollbacks President Bush has done through regulation, catching up on the environment isn't going to happen anytime soon," he replied.

I gazed across Lynn Canal at the sharp-edged slopes of the Chilkat Range. "Come on, Bill! I want to believe again. Go with me here. It's the New Year and… and… holy shit, its two whales breaching!" Bill, who was fidgeting with his pack, looked up, but only in time to see the huge splash. We stood transfixed, silenced by the spectacle of a pair of humpback whales playfully and repeatedly slapping the water with their large (at least 10 feet in length) white pectoral fins.

While seeing humpback whales is common in summer in Southeast Alaska, it is an unusual sight for winter, when most of the humpbacks migrate to Hawaii. The sight of two whales breaching together was a first for me, a longtime resident who has had the privilege of a whale diving directly under my kayak. I walked along in silence, re-creating the image in my mind. Both whales were facing the same direction; they were synchronized. My mind spun with the significance of the event. Still basking in broken light, my heart swelled with the hope and promise of a new year.

That evening, I gathered with friends for the annual New Year's campfire on the beach. While roasting hot dogs, I relayed the day's experience. After a round of head nodding, one of my friends said, "It's just like in Native lore where you mark the seasons or the year by the sighting of unusual events, like the *Summer of Three Bears Swimming*." Raising a paper cup of champagne, another friend chimed in, "Here's to the Year of Two Whales Breaching." Raising my cup, I cheered, "Here's to the Year of Two Whales Breaching, the year the political pendulum swings back."

I held tightly to the power of this imagery, occasionally telling the story when a conversation turned downward into a spiral of negativity. I mostly kept my hope story to myself, not wanting to put too much stock in the changing of the political winds, not wanting to set myself up again for the disappointment of 2004 when George Bush won a second term, despite the country knowing that the "weapons of mass destruction" had never been found. Never mind that the Iraq war was spinning out of control.

Eleven months later, the mid-term elections of 2006 sent a wave of reform from the East Coast to the West Coast. Secretary of Defense Rumsfeld resigned, signaling a new direction in our Iraq policy. There was talk of middle class tax relief. In Alaska, the old guard Governor, Frank Murkowski, was tossed out of office and a coalition of Republicans and Democrats took over leadership of the Alaska legislature. Most importantly for the whales and for the nation, politicians on both sides of the aisle began talking about climate change legislation. According to news sources in Washington, oil executives were leading the charge for a cap and trade program on greenhouse gases. Even though a cap and trade program allows for trading pollution credits instead of banning them, the model had success with reducing sulfur in the atmosphere and consequently reducing acid rain. It held promise as a model on how to use the marketplace to regulate carbon.

With news about a carbon cap and trade deal in the works, I was delighted. With this tidal change in the news, I believed again. I wrote about the year of two whales breaching and shared it with friends.

Ten years later, I can only laugh at the naiveté of my optimism. Since 2006, the oil executives that once supported cap and trade legislation have abandoned the deal, and instead put their financial resources behind a Koch Brother's campaign to create climate doubt.[3] Adding injury to insult I later learned the cap and trade deal that the environmental community was once all excited about and even touted as historic, handed out enough free allowances to let ninety percent of emissions from energy utilities, including coal plants go unregulated.[4] I didn't think the national green groups would sign onto something so weak. I didn't think the oil companies that use climate science in their operational plans would not only walk away from a climate deal but ultimately the science itself.

Then in January of 2010 came the Supreme Court decision in *Citizens United* by which the Supreme Court gave corporations and unions the green light to spend unlimited sums on ads and other political tools, to elect or defeat individual candidates. This ruling combined with other court rulings recognizing corporations as "persons" led to the creation of what is known as Super PACs (Political Action Committees). These committees then allowed small groups of wealthy donors to gain even more influence over elections than was previously possible. And because elections have become ever more expensive these wealthy individuals and corporations have since been able to maintain considerable influence once a candidate takes office. Unfortunately, many of these individuals and corporations having accrued their wealth from extractive non-renewable resources now exert their rise in influence to thwart progress of climate change legislation. I didn't see any of this coming when I mused about the significance of two whales breaching in godlight.

But there has been another significant political development since that day, one that gives me hope, one that I also did not see coming. In 2008, the millennial generation (those born after 1982) showed up in national elections and made a huge splash, securing the election of the first African-American president. Given the role of young people in the rise of Senator Bernie Sanders' campaign, it's clear we are still riding the impact waves of this generation. And we are still watching for whales.

Every spring my neighbors and I are on the lookout for killer whales (also known as orcas) that wander into Gastineau Channel scouting for herring. Gastineau Channel is an eleven-mile fjord right outside my kitchen window. In Native lore, killer whales symbolize family, harmony, travel, and protection. They are said to protect those who travel away from home and lead them back when the time comes. Known as the "Lord of the Ocean," the killer whale is said to be its guardian, with seals as his slaves and porpoises as his warriors.

So when the guardian of the ocean and his family came up Gastineau Channel one day, the phones began to ring. "Orcas in the Channel!" said an excited neighbor. Gastineau channel ends in a mudflat only two miles above my house, so the orcas would have to make a U-turn and pass me again as they returned to the deeper waters of Stephens Passage. Before long, the friends and neighbors of Douglas (a neighborhood area of the capital city of Juneau) were out on the harbor breakwater to get a closer look.

Realizing I still had time before the orcas came back through, I hurried to strap my kayak on my car and raced down to the harbor to launch it. Ever respectful of their protective pod formation, I hung close to shore watching for their return. In short order, I spotted their spouts and dorsal fins in the distance. The drifting currents brought me toward the middle of the channel.

The pod of eight was now in clear sight with two massive males leading. Coming into the narrowest part of the channel—just ahead of me—the males began to veer toward shore. I quickly back-paddled a few strokes. I announced my presence with a sharp slap of my paddle against the bow. My heart raced. The rest of the pod caught up mid-channel. Suddenly, the alpha male changed direction. I panted in relief and managed to get a couple of close-up photos before returning to shore.

At home, I decided to use the adrenaline rush of my orca encounter to write my next op-ed. I resolved not to hold back in my critique of former Governor Parnell's dismantling of coastal management in Alaska. Despite 44,000 miles of coastline, more than all the lower 48 states combined, and a coastal program supported by municipalities and communities throughout Alaska, former Governor Parnell did the bidding of Big Oil and killed the program in 2011.[5] Alas, Alaska is now the only coastal state in the United States without a coastal management program. As you can see, the School of Hard Knocks continues—but so do the wonders of whales.

Chapter 2 — In the Time of Glaciers

(This essay was first published in the Oxford Journal, Interdisciplinary Studies in Literature and Environment, Volume 21, Issue 1, Winter 2014)

Figure 2: This twenty-six year time lapse photo shows the massive amount of retreat and ice loss of the Mendenhall Glacier. Photo courtesy of award winning photographer Mark Kelley, © Mark Kelley, Juneau, AK 2012.

Whenever big questions tug at me, I find myself returning to the Mendenhall Glacier, my marker of climate change. The two camping photos above were taken twenty-six years apart from the alpine meadow of Thunder Mountain. The photo on the left was taken in 2014 and by the contrast shows a glacier that has melted back more than a mile and shrunk by 700 vertical feet.[6] Although the glacier has been retreating since the mid-1700's the rate of retreat has been rapidly advancing. In the last thirty years it has retreated more than the previous one hundred.

As a winter playground, the glacier becomes ever present in my life for half a year. Not wanting to miss the one or two days that solid ice and sunshine collide, my friends and I are sharing scouting reports. One of our favorite things about living in Juneau is the opportunity to cross country ski on Mendenhall Lake in glorious sunshine. Your kick and glide is lifted by the sparkling snow. In the distance lies a glacier capped by seven granite spires. Sprinkle in the sweetness of friends and it's a truly magical day. And you forget completely about climate change.

This is the same when a warm spring brings kayaking on the lake in May. A stiff breeze blows off the glacier. The grey, spitting clouds are not the least bit inviting. Even so, I paddle my kayak across the lake to the base of the towering wall of glacier ice. The deep blue colors of the ice lure me closer, while the moans of cracking, heavy ice caution distance and respect. It is the tug of life carried out in one of nature's grandest settings, the Mendenhall Glacier, which descends twelve miles through seven rock towers to release icebergs into the lake.

I'm drawn here today not just by the alluring blue, but in the hopes of seeing Arctic terns returning from Antarctica. After journeying more than 10,000 miles from the other end of the world, terns roost on the grounded icebergs. Their marathon journey ensures that each bird sees two summers each year and more daylight than any other creature on the planet. In other words, more than any other species, the arctic tern seeks to live its life in light. This is something my colleague, Terry Tempest Williams first pointed out to me when visiting Juneau.

I've been visiting the Mendenhall Glacier regularly for the last quarter century. Through those years I've come to think of it as part of my everyday life, a friend of sorts. Like any friend whom one visits repeatedly over the years, the Mendenhall Glacier marks my family memories. Two decades ago, I played on the sandy moraine with my school-age children. Then the face of the glacier stood near the waterfall. Now it is a half-mile away and my children are grown.

Not only have I played here; I've prayed here as well. I affectionately call the Mendenhall the *Church of the Latter Day Glacier*. Listening to a deep blue spear of ice groaning to fall free from the glacier's hold, I've prayed for loved ones to be released from terminal illness. Here, I can understand the transformation from ice to water, from river to ocean, from life to death and life again. The glacier, in its largest meaning, comforts me.

Now, on exceptional days, it renews me. When the low sun stretches a beam of heat down the length of the lake where friends are skiing on a fresh dusting of snow, it warms the child inside of me. My favorite activity on these gift days is ice skating. Nothing makes my spirits soar more than skating effortlessly around icebergs whose magical shapes and hues of blue somehow match the power of the music thundering through my headphones: Beethoven's *Ode to Joy*. It's an explosion of beauty, power, and creativity. On these days my spirit is uplifted to carry on the good fight.

In 2012 the United States experienced the warmest year in 118 years, which is as far back as records go.[7] Hurricane Sandy walloped New Jersey and New York with more than $80 billion in damages.[8] The cost for wildland firefighting has grown from $100 million in 1960 to $3.3 billion today. The summer of 2016 drought in the Midwest resulted in a loss of $32 billion in crop failures.[9] There is now no part of the United States untouched by the economic impacts of climate change.

The Mendenhall glacier continues to retreat at a rate no one expected. Soon, there will be no more icebergs falling into the lake. Though a glacier's decline sends a strong signal, we all know that climate change is so much more than this. It is a disruption of ecological balance at a magnitude greater than we ever thought. Biodiversity is plummeting as the Arctic melts, which threatens an entire biome of ice-dependent species. The earth's forests are burning more and burning hotter.

The world races blindly toward the 2° Celsius maximum increase in global temperatures, a relentlessly approaching benchmark set by forty-two industrialized and forty-three developing countries. Scientists worldwide established the level as a warning to avoid irreversible, out of control effects of climate change at that temperature benchmark.

According to the *World People's Conference on Climate and the Rights of Mother Earth*, a U.N. sponsored conference held in Bolivia on Earth Day, 2010, exceeding the 2 degrees Celsius "results in a fifty percent probability that the damages caused to our Mother Earth will be completely irreversible. Between twenty percent and thirty percent of species would be in danger of disappearing. Large extensions of forest would be affected; droughts and floods would affect all regions of the planet. Many island states would disappear and the production of food would diminish and the number of people in the world suffering from hunger would increase dramatically, a figure that already exceeds 1 billion people."[10]

Yet we remain on a path to exceed this critical benchmark. According to the United Nations Panel on Climate Change (IPCC) the world's "carbon budget," the amount of greenhouse gas that can be emitted without exceeding 2° Celsius, could be used up entirely by 2040.[11]

President Obama pledged a 26-28 percent reduction in greenhouse gases by 2025.[12] This became a heavy lift without Congressional action and is threatened by the election results of 2016. Tragically, Congress remains entirely inept, not only on climate change, but on every fundamental problem facing us. Now, thanks to the Supreme Court's ruling in *Citizens United*, the odds of getting federal climate change legislation passed have worsened from daunting to nearly impossible.

It is one of the greatest ironies of our century: never was there a time when we were collectively so smart about understanding future trends, yet acted in ways so utterly and profoundly stupid.

This is what the current generation will be thinking about my generation when, after decades of climate-driven disasters, the second half of the century arrives. This is a disturbingly sad legacy to leave. I feel an immense sense of guilt. Even though many of us have been awake and active, some out in front, others from behind the scenes, all planting seeds for members of the current generation to build on, it's hard not to feel that my generation has failed the younger generations in our mission to leave behind a healing Earth. After all, we are baby boomers christened by the activist politics of Vietnam, civil rights, and the women's movement. We are trying to address climate change, but we can only do what was doable in the times of climate denial. As one among the thousands of my generation's activists, let me apologize for our collective inability to achieve ground-shifting policy changes.

As my backyard glacier shrinks, becoming weaker, showing more bare rock, it's hard to face this bold, barren barometer of climate change. Now I place my hope in young people. I have high hopes that the millennial generation will rise to become the much-needed civic generation they are touted to become by sociologists. After hearing an NPR interview about the shifting political values of the millennial generation, I was intrigued to read and learn more.

Historians Marley Winograd and Michael Hais who wrote *Millennial Momentum*, a fascinating exploration into America's historical demographic, "About every eight decades, coincident with the most stressful and perilous events in U.S. history—the Revolutionary and Civil Wars and the Great Depression and World War II—a new, positive, accomplished, and group-oriented 'civic generation' emerges to change the course of history and remake America. The millennial generation is America's newest civic generation."[13]

History apparently shows that members of the civic generations (the last one was the GI generation) focus on resolving societal challenges and building institutions. Winograd and Hais explain that: "Despite coming of age during periods of intense stress and turmoil, civic generations invariably exhibit a uniquely high degree of optimism about where they and the nation are ultimately headed." Using these patterns of generational archetypes, the authors analyze current economic and social trends to project changes in the way Americans will work, learn, govern and live. In regard to how these trends play out for saving the planet, they project that the "millennial enthusiasm for taking individual action to fulfill collective responsibilities in the marketplace, as well as in the voting booth, *will* (emphasis added) tip the scales in favor of pro-green policies at both the national and international level."[14]

I added the emphasis to the word "will" because the young people I have worked with when serving as the Executive Director of Alaska Conservation Voters bear this out. They were the ones designing a new conservation scorecard for evaluating state legislators and setting up phone banks to support pro-conservation candidates. They were the ones who volunteered to work the phones over the weekends and canvass key neighborhoods at the last minute.

Although I believe my long-distance hope is well placed in the millennial generation, I want to leave lesson landmarks to help along the way. I know the winds of climate change will be blowing long and hard. I know there are barriers to sustainability. In Part Two, I mix policy beacons in with my personal and professional experience to establish a few helpful markers for the journey ahead. There are lessons to pass on to those civic minded millennials, the ones on my mind as I paddle toward that barren barometer of rapid change, the Mendenhall Glacier.

Searching for Arctic Terns and with a full heart, I circle the first set of grounded icebergs. The tip of my kayak nudges along the brow of a large berg. No terns roosting. Realizing that these bergs are within sight of the Visitor Center, I remain hopeful and paddle toward the larger, more distant icebergs.

As the wind picks up, I seek the lee of a house-sized berg. Tucked inside, there they are—a row of four mega-marathon champions, decked out in black caps and sharp-looking tail feathers, all sitting on a ledge of ice. Ever proud of their arrival, the terns are hope on long-distance wings. The terns take flight as I round the corner, displaying aerobatic maneuvers that the best of pilots can only dream about. Hope remains at the calving edge of the Mendenhall glacier, my marker of an ever-changing world.

PART TWO   ADVENTURES FAR AND DEAR

Chapter 3 — In All Places Green

Figure 3: IES exchange students attending Durham University, 1972-73. The class photo is taken in front of Durham Castle. Front and center is Professor Spooner, a man of intellect and humor who "sort of" chaperoned us. Kate is in the second row, fourth from the right.

In 1972, the women's movement in America was in full swing and the country was deeply immersed in the Watergate hearings. That's when I headed off for England. I was one of forty-four American students from colleges all across the states selected by the Institute of European Studies to attend Durham University in the north of England near Newcastle and Scotland. It was my junior year abroad and instead of going back to school in Iowa, I was going to a land of castles, cathedrals, and pubs.

Durham is best known for its eleventh-century cathedral and castle on the banks of the River Wear, the home of Durham University since 1832. Both the cathedral and the castle are now UNESCO World Heritage Sites. I remember first walking inside this massive Norman cathedral and feeling instantly transported to another world. And indeed I was.

Durham University is one of three collegiate universities in England—Oxford and Cambridge being the other two. At a collegiate university, the dining hall, pub, dormitory and library are all in one college, and a university is made up a of dozen or so colleges. Colleges are much more about belonging to a community, with a common identity, traditions, sports teams, societies and activities. Accommodation is just one aspect of being a member of a college. I was at Trevelyan College while my boyfriend attended University College housed within the castle confines. Within the college I had a moral tutor. He poured the sherry and I talked. Next was discovering pubs for the first time. Oh, I was so very far away from Iowa.

My favorite part of English college life was the formal meals twice a week with the entire college. At all formal dinners we wore black academic gowns and stood when the high table entered the dining hall. Fortunately for me, my sort of boyfriend, John, was assigned to University College, so his formal meals were served inside the Great Hall of Durham Castle. Fortunately, having guests from other colleges within Durham was permitted. Whenever I got invited to dinner there we started off in a tiny pub in the castle basement called the Undercroft. Then on up into the Great Hall where the transformation to the eleventh century became complete. The Great Hall was once used to wine and dine kings, clergy and noblemen throughout the north lands. Then the three walls of stained glass take your breath away.

This transformative effect of The Great Hall is one that many readers may have experienced. Remember the *Harry Potter* dining hall scenes with Dumbledore sitting below a flaming chandelier? Yep. Durham Castle. But of course, it's on a whole other level with the *Harry Potter* series.

Although I did not encounter any witchcraft classes, the year I spent in England and hitchhiking around Europe stands out as one of the most magical, important times in my life. Leaving behind a tumultuous United States still in the throes of the Vietnam War and Richard Nixon's devious Watergate legacy, we were idealistic students out to discover our place in the world.

On the weekend breaks, I often hiked in England's renowned romantic heart, the Lake District. As a reader of Wordsworth, Keats, and Coleridge, I fell in love with the garden-strewn lakeside villages surrounded by mountains and lush pastures lined with stone fences. It was all so tidy and green. Gurgling brooks greeted me as I wandered along the pathways of intricately laid stone; stone placed at angles meant to be appreciated for hundreds of years. It was my first encounter with a sense of harmony—the village had a seamless encounter with the lush green surrounding hills. It was here where I discovered my favorite Wordsworth poem, "Lines Written above Tintern Abbey," a poem that has stayed with me all my life.

> *To look on nature, not as in the hour*
> *Of thoughtless youth; but hearing oftentimes*
> *The still sad music of humanity,*
> *Nor harsh nor grating, though of ample power*
> *To chasten and subdue.—And I have felt*
> *A presence that disturbs me with the joy*
> *Of elevated thoughts; a sense sublime*
> *Of something far more deeply interfused,*
> *Whose dwelling is the light of setting suns,*
> *And the round ocean and the living air,*

And the blue sky, and in the mind of man:
A motion and a spirit, that impels
All thinking things, all objects of all thought,
And rolls through all things. Therefore am I still
A lover of the meadows and the woods
And mountains; and of all that we behold
From this green earth; of all the mighty world[15]

The Lake District is where my love affair with nature took off. So it's in tribute to England that I start this mini tour of places green. Given the many years of absence of U.S. leadership since the Kyoto Protocol, it's imperative to build on the Brits, the Danes, and those countries that are leading the world in transforming energy and building sustainable communities.

In this regard, it's good to know that England in 2016 ranked twelfth out of one hundred and eighty countries in overall environmental performance as determined by Yale University's Environmental Performance Index (EPI).[16] By comparison, the United States placed twenty-sixth.

The Environmental Performance Index (EPI) ranks how well countries handle high-priority environmental issues in two broad policy areas: 1) protection of human health from environmental harm, and 2) protection of ecosystems. England is high on the list due to their attention to sanitation and access to clean water. Furthermore, England passed a climate change act in 2008 and has stepped up their commitment to emission reduction under the Kyoto Protocol.

My first big adventure out of England was on our Christmas break. As the English students at Durham went home for the holidays, we, the Americans, set off for adventures in Northern Europe. Some of the adventures were as a group because the first two weeks were all arranged by the Institute of European Studies. After the first two weeks, each student was on their own to get back to Durham by the first day of spring semester.

The first European capital on our group tour was Copenhagen, Denmark. It was here that I learned to appreciate specialty museums and how an engaging guide can make history come alive. To illustrate the transition, the well-informed guide said, pointing to the Late Greek's mouth, "Look, here are lips that can kiss." Moving on to the earlier period, "imagine these lips kissing"? I could dramatically see the difference in style after that gem of insight. Admiring his zeal I hoped that someday I too could do the same in whatever career came my way.

Now, forty-three years later, my hopes are less personal and more global. For example, now when I think about Denmark and paths, I hope that someday my country will find the same zeal in tackling climate change.

Denmark is a country pursuing the world's most ambitious climate change policy to date: to be fossil fuel free by 2050.[17] The Danes have "essentially invented the modern wind-power industry." An article in the New York Times notes that the Danes are above 40 percent renewable power on their electric grid, aiming toward 50 percent by 2020. With their expansive district heating program and their push for electrification of vehicles, Denmark is well positioned to meet their goal. And they intend on doing it with a strong economy.

The Yale University Environmental Performance Index results break the country score down so you can see how each country ranks on climate and energy, and whether or not governments are moving in the right direction through their laws and policies. According to Yale's 2014 EPI report, 43 percent of the world's high and middle income countries are reversing their trend in carbon emissions. These countries, like Denmark, are slowing down the rate of carbon emission and moving in the right direction.

The top twenty overall environmental performing countries are listed in Appendix A. When using this EPI ranking, it's important to know that each metric does in fact send a policy signal and that competition among nations drives a race toward the top, spurring governments to improve their environmental performance.

When you hit the *Climate and Energy* tab on the EPI report, England drops out of the top twenty countries, but Ireland rises to number twelve (2016) in the world due to lowering its trend in carbon intensity. What I find most inspiring about Ireland's green achievements is their Framework for Sustainable Development, signed into law in 2012. It sets a long-term, detailed structure for advancing the green economy in Ireland.

Ireland not only stirs my professional admiration, but also my blood. I am about ninety-five percent Irish-American (the five percent not Irish is where my name comes from). So while abroad as an exchange student, I was certainly going to check out Ireland on my spring break.

It is here in Ireland, circa 1973, where I described myself as a "pragmatic romanticist." In re-reading travel journals from when I was twenty, I find it amusing to note that one minute I'm feeling transcendent by my pastoral walks and poetry readings, and the next minute I am all about politics, engaging in conversation about the role my generation will play in addressing the global issues of the environment, racial justice, and peace. The setting for these intense conversations was, of course, Irish pubs.

I was in the village of Crookhaven, within County Cork, Ireland. I was thrilled to be there on many levels. First and foremost, Ireland is the land of my ancestors where stories and legends color the raw character of the land and sea. Secondly, it was the first chance I had to be out into remote countryside since leaving the bustle of Rome. The stone-lined village of Crookhaven sits along the coast of a peninsula known as Mizen Head and was slowly coming alive with holiday returnees.

My good-natured friend, big blond Patrick (nothing between us) and I kept bumping into two cheerful middle-aged ladies who came from England to Crookhaven every spring for the flowers. We had such nice conversations whenever we ran into these opinionated, humorous ladies, so why not meet up at the Crookhaven Inn?

Apparently the directness of my idealism made for good teasing; enough for me to write about.

A couple of gin and tonics and the corresponding conversation grew intense. The witty, loud English lady purposely challenged my optimism about the "young kids today" righting the wrongs and bringing the world closer together through mutual understanding.

"Mind you, we've got a big task but if there's any generation suited for the job, we are," I told her. I also got to give her my environmental talk about figuring out the carrying capacity of the land around us, of the Earth.

Finally, she taunted me with the threat that once I got to her age I would end up copping out just like her; that I too wanted a cozy house tucked away from it all to read great books.

Back in 1972, the challenges were cleaning polluted rivers, getting rid of harmful smog, protecting old neighborhoods from highway expansions. Now it's climate, biodiversity, population growth, and saving the rainforest. We now live in a world where we know that the rate of deforestation in Brazil matters to a quaint Irish village on the coast, a village that will feel the effects of sea level rise and ocean acidification as climate change accelerates. Now I am the age of the English lady that chastised me. I too live in a cozy house overlooking the sea. But I am not "tucked away from it all."

Nor is the millennial generation tucked away. They too are pragmatic idealists by nature. In his ground-breaking book, *Fast Future–How the Millennial Generation is Shaping Our World,* millennial David Burstein devotes an entire chapter to why young people view themselves as "pragmatic idealists." Burstein, a commentator on youth and politics for a range of media outlets, notes the difference between the activism of the baby boomer generation and the activism of the millennial generation:

While there were moments in the 60s and in other eras when pragmatism and idealism were combined and balanced, they have not been fused together in the prevailing mindset within a rising generation until now. Today's millennials generally view change in society as a project to work on, not something to demand. We know that, in order to effect change on issues we care about, we have to master the workings of our society's existing institutions.[18]

The millennial generation realizes that the most complex problems like climate change require a meaningful mix of idealistic vision with a pragmatic approach. Burstein assures us, "We are not passive on these pressing issues like sustainability; we are simply learning to work within the system. We are aiming for change on a larger scale."

Think about the global scale of climate change and then read these words again. Is this not hope on long-distance wings?

I also see hope in the fact that more and more young people are studying abroad. Being worldly at a young age opens eyes and hearts. According to the CBS news program *Moneywatch,* the number of American college students studying abroad is at an all-time high. Most of them (55 percent) study in European countries with the United Kingdom the most popular destination.[19] Hmm, some things don't change.

In looking at the top 10 destinations list I wondered how these countries matched up on Yale's top ten EPI rating. Only Spain and France overlapped on the top ten of both lists. England, Australia, and Ireland also scored well on both lists. Students going to these countries could see environmental models to emulate back home and help get America engaged on the same level.

French President Nicolas Sarkozy says, "You [America] cannot be the first champion of human rights and the last when it comes to obligations and responsibilities of the environment." [20] Germany's federal environment minister, Sigmar Gabriel, agrees. "America is the most dynamic country in the world and the biggest economy in the world. We need American capitalism applied to this problem…If the Americans are going green, the whole rest of the world is going green."[21]

Making America greener seems to be the perfect type of project for the "pragmatic idealist" generation. If I were sitting in a pub with some college-age students from America, this would be the challenge I would taunt them with.

I visited Germany as well (number seven on the popular study abroad list). Here too I discovered castles, as Kathleen (my hitchhiking buddy) and I travelled down the Rhine. Once in the Middle Rhine region we headed out to explore the largest medieval castle overlooking the Rhine. Our rides took us to the village of Oberwessel, the closest we could get to the castle. Here's one last castle passage:

After a brisk 8 kilometer walk to Sankt Goar, we immediately headed up to the castle. It was closed. That did not stop us. We came too far so it was over the fence we went. As we literally crept around the castle to keep out of sight of passing cars we found the dark tunnels and passageways exciting. I felt as if I was living a childhood fantasy of exploring a mysterious castle essentially alone. After hopping back over the fence I felt immense relief of not getting caught and it was this sense of relief that made this castle exploring particularly spectacular. Something that would not have happened had we not jumped the fence.

Being worldly at the impressionistic age of twenty-one gave me wings of confidence. I share this last journal tidbit because I think it's important to be willing to occasionally jump the fence as one pursues professional goals. Being worldly and willing to jump fences on occasion— wings with an updraft every now and then.

Now I just needed to find roots to go along with my wings.

Chapter 4—What Good is an Eagle if You Can't Eat It?

Figure 4: "Eagle Bait", colored pencil drawing courtesy of Ray Troll, © 1988, *Ketchikan, AK.*

I arrived in Alaska in 1977 from Socorro, New Mexico, where Bill and I had been working as range technicians for the Bureau of Land Management (BLM). We had just earned our Masters degrees from Yale School of Forestry and Environmental Studies, and Socorro was where we landed for summer jobs, basically counting cows on ranches with leased BLM lands. By the end of summer we were ready to head out to be near big trees or big mountains. This meant either moving to the Pacific Northwest or Alaska or, more specifically, wherever one of us could find a job. Being ever responsible and organized, I had sent out at least one hundred resumes while Bill had done nada, zip, save keeping current on a federal job list. Naturally he was the one who got the call, out of the blue, to work as a land law examiner for the BLM in Anchorage. It was a federal paper-pushing job to convey lands to Native Corporations under the Alaska Native Claims Settlement Act. He was needed right away for training. Could he be there in a week's time?

After answering yes, he turned to me and asked, "Are you ready to move to Alaska? Oh yeah, and by the way, why not get married?"

Comforted by the fact that I had a long solo drive ahead with plenty of time to think about the big decisions, I said "yes." I figured if I still wanted to marry Bill by the time I got to Anchorage, Alaska over 3,500 miles away, he would be the right man for me. If the notion of getting hitched made it through Joni Mitchell followed by Aretha Franklin for the long mountain climbs and then of course Janis Joplin on the long, boring stretches of road, he would be good to go. No turning back.

As the drive progressed into week two, I realized that deciding to move all this way to Alaska was the big decision; getting married to Bill was secondary. If Alaska felt right, then Bill was right. That simple.

After driving to Anchorage, it took me a year to land my first professional gig as a resource planner with a consulting firm. The firm had just won the bid to draft and develop the Haines Coastal Management Plan, the first coastal management plan in the state. I was excited. Here was a prize opportunity to get in on the ground level of doing things better in Alaska.

We could learn from the resource management mistakes of the Lower Forty-eight and start off by doing things right. What a perfect place to begin, Haines.

The community of Haines lies at the northern end of Southeast Alaska, the jumping off point for the Alaska Highway to Anchorage some eight hundred miles away. The granite spires of the Cathedral Peaks frame this town of 2,500 people into a Yosemite-like setting. On a clear day, this is one of those rugged Alaskan views that takes your breath away. Better to number your life by the number of places that take your breath away than the number of breaths you take. Or so says the foreword to Patricia Schultz's book, *1,000 Places to See before You Die*. The view of Haines had me gasping as I sought to do right by this beauty. The perfect place to begin my career.

Or so I thought.

The resource issues in Haines were the types of conflicts that many small towns across rural America experience. In 1979, Haines was a timber town with a closed sawmill. The Chamber of Commerce blamed log shortages and EPA permitting rules. While a few town officials talked up the mission of economic diversification through fish processing and tourism, most movers and shakers acted more like screamers and railers, demanding "*more wood*". It didn't take long before environmental bashing dominated the community sentiment.

The conflict boiled down to one issue: either you supported the designation of all state land for logging or you supported "locking up" land for the Haines Bald Eagle Preserve. Each autumn, over three thousand bald eagles converge along the Chilkat River to feed on a late run of Coho salmon. It's the world's largest concentration of bald eagles. Since those officials touting economic diversification had no concrete plan, environmental bashing held sway and any mention of the eagle preserve relegated one to the lowly status of tree hugger. To those living in red states, does this conflict and name-calling sound familiar?

Even knowing these community dynamics, I found myself completely caught off-guard in my first public meeting as the city's planning consultant. I thought I was attending a preliminary meeting with the City Council to go over the scope of services for developing a coastal management plan. Instead, I found myself in a no-holds-barred public meeting about how coastal management related to the eagles and tree debate. In the midst of my initial explanation of the planning process, a scruffy-bearded logger wearing his red-checkered jacket rose up and interrupted me with a question.

"What good is an eagle if you can't eat it?" he asked sternly.

"I'd like to hold that question for a few minutes until I lay out the scope of work as the city manager requested," I said, using my best unattached professional voice.

Dead silence. Rustling in chairs ensued and the logger rose out of his chair again. "Enough of this planning bullshit. What good is an eagle if you can't eat it?"

Turning to the city manager for direction all I received was a short nod to proceed. Now my gut was screaming. *What type of absurd question is this? Eagles are magnificent birds, a sight to behold, our Nation's symbol of freedom and patriotism, and part of the heritage of the native people. You want to eat an eagle? Are you nuts?*

"Well, for one thing, eagles eat all the rotting salmon in the river. Parts of town would smell a lot worse without them," I said, searching for time to think.

"So do bears, but I can eat them."

Now the room shook with hoots and laughter. Buying me more time, the old logger went around the room enjoying a round of backslapping at my expense. At last the room quieted down.

"If you really do want to diversify your economy," I told them, "you have to use all your assets. You have the world's largest concentration of bald eagles. This is what sets you apart from other Southeast communities that compete for tourism traffic. Because you have the native village of Klukwan nearby, you could compete against Ketchikan for visitors who want to learn more about native culture, but Saxman already has a totem park nearby. You have glaciers like Juneau, but none that you can drive to. Like Skagway, you have a link to the gold rush. You can try and compete or you can carve out a different niche. Haines can capitalize on what no other port has: a world-class natural history phenomenon that any patriotic American can relate to. You may not be able to eat them, but if you work it right, the eagles just may be your most immediate ticket for new jobs that put food on the table."

"Okay then," said the logger. "I'll sit down now that you've finally answered my question."

When I finished with a few examples about an eagle visitor center and Audubon tours, a low buzz of conversation and head-nodding undulated through the room.

When I had time to reflect on the dynamics of the meeting, I realized that if I had justified the land allocation for the eagles out of the eagle's intrinsic natural value, I would have bombed royally as a city consultant. By portraying the eagle as an economic asset, the discussion shifted from eagles *versus* timber to eagles *and* timber. I had experienced one of those paradigm shifts that you hear about at professional conferences. Instead of viewing possible solutions to resource conflicts as a search for a reasonably balanced plan, I should search for the synergy between environmental and economic concerns. Instead of looking for trade-offs, I should be looking for ways to fuse agendas, to translate environmental concerns into economic matters and vice-versa. If I found the linkages, then hopefully the balanced solution would come naturally.

In the thirty-five years since Haines, I have widely expanded my resource management and political horizons. I have experienced many such "linkage moments." My perspective on these moments has eventually crystallized into a working philosophy that I call "*eco-nomics.*"[22]

Eco-nomics is now a term applied more broadly to encompass the field of environmental economics. The synergy between economics and ecology also brings forth this truism: *it's impossible to have a sound economy without a healthy environment.* Stability of one system promotes stability in the other system. The trick is to get decision makers to recognize this and act on promoting both at the same time. Not easy. Therefore, eco-nomics is also about techniques and insights gained from my linkage moments.

The how-to is explained through the use of nine eco-nomic principles that can be found in Appendix B. Following these principles puts us on the path of changing the jobs *versus* environment scenario into a jobs *and* the environment scenario.[23] Here is a sample principle—Synergy Rules.

Eco-nomic Principle 8: Synergy Rules

In a fully evolved ecosystem, waste and pollution do not exist. What is waste to one species is often food for another. In a fully evolved economic system, one company's waste is another company's raw product material. Companies paying attention to the green line—recycling, green marketing, waste reduction, and energy efficiency—are more stable and profitable in the long run than those that ignore environmental practices. Proper use of the marketplace can accelerate the benefits back to the forest, mountains and oceans. Man, at the scale of seven billion people, and Nature are forever entwined. We all must recognize that progress and prosperity for humankind, if done wisely and compassionately, can be good for the environment as well.[24]

These nine principles still apply to today's challenge of promoting sustainable use of renewable resources, but their application is compounded by the rise of corporate influence in our political affairs. Back when I was twenty-seven years old, I only had to deal with a dominant timber industry in a small town. Now, anyone working on climate change must deal with large international oil companies whose influence has been flamed by the massively errant decision made by the Supreme Court in *Citizens United.* I fully recognize that the bar is higher.

What became of the Haines' timber and bald eagle battle? The plan recommended the establishment of both a State Forest and a Bald Eagle Preserve. The plan was approved by the City Council in 1979 and became the first coastal management plan approved in Alaska. In 1982 the State of Alaska established both the Haines State Forest and the Chilkat Bald Eagle Preserve. If you travel to Haines today, you will not see the old timber mill. In the end, the market was a greater problem than supply. Instead of the mill, Haines has re-branded itself as the *Adventure Capitol of Alaska*. Cruise ships and a diverse cadre of tours serve a tourist industry. In November, Haines hosts the annual Alaska Bald Eagle Festival. Visitors from around the world come to partake in the fascinating migratory event of 3,500 eagles converging on the beautiful Chilkat River Valley.

Best of all, the heir to the timber mill is now on the Haines City Council, spearheading an effort to build a wood pellet mill to displace oil in heating local buildings. She has the full support of the community in this endeavor.

Chapter 5 — The Way of the Earth's Elders

Figure 5: "Alaskan Gothic", acrylic on canvas painting courtesy of Tim Troll, Anchorage, AK, 1992.

Wandering around the dirt streets of Togiak, I felt a country away from the community of Haines. Although within the same state, Togiak is a remote traditional Yup'ik Eskimo village, nine hundred air miles away into the tundra of Southwest Alaska.

My next professional endeavor with the Alaska Coastal Management Program had me exploring the outskirts of an enclave of half-painted plywood homes. In 1981, over 200 Yup'ik Eskimo lived a subsistence lifestyle along a sweeping wide river. Walking down by the barge landing, I noticed a handmade flyer on a telephone post. In red marker on cardboard was a notice about a Yup'ik dance lesson at the school. It started in an hour or thereabouts.

Along with another woman, I found the school gym door. In the foyer were six Yup'ik grandmothers teaching skin sewing to a small gathering of teachers and students. Upon seeing us, they opened their circle to show us what was for sale on their table. One of the teachers translated for us as we inquired about buying seal mittens and zipper pulls. We asked about baskets. Out came a box of woven grass baskets. Then the oldest, most round-face, crinkled elder reached below and brought up a box of rocks. She was grinning ear to ear. With a rock hound for a husband, this immediately caught my attention. I dug down into the box and pulled up a broken piece of spherical stone. It appeared to be fossilized bone. The sight of it brought on a round of giggles.

"*Oosik*?" I queried. More wide, tooth-missing grins popped out leading to squeals of laughter. Knowing that an oosik was the penis bone of a walrus, I mimicked being grossed out by the bone and dropped the bone back in the box. I picked up another oosik and held the more intact, longer oosik aloft. The grandmas doubled over with laugher. I was in stitches.

I knew then that my new job with Bristol Bay Native Association (BBNA) was going to be like none other. Where else did you need a translator within your own state; within your own country for that matter?

My job as Coastal Zone Management Director was to travel to all thirty-one Yup'ik, Aleut, and Athabaskan villages within the service region of BBNA to educate them about the local governance benefits of participating in the Alaska Coastal Management Program (ACMP). In many Yup'ik villages English was hardly spoken at all and the deeply guttural language was like being back in Yugoslavia (spring break trip while at Durham). As Alaska is large so too is the entire Bristol Bay region (62,500 square miles) and by comparison roughly covers the same amount of land and water as does the entire country of Greece.

My job was to explain a state resource management program connected to the federal government. If they did a management plan they would have a say—not a guarantee—in resource decisions made by state and federal agencies. But first they needed, through an election, to establish a government area, the Bristol Bay Coastal Resource Service Area. I was there to collect on behalf of BBNA, supporting resolutions from a majority of Village Councils within the Bristol Bay region.

Although I did not need an interpreter for every village, I did need to bring to each village a healthy appetite for subsistence foods. In Yup'ik villages I was treated to beaver and seal, while in the Aleut communities, crab reigned. The two Athabaskan villages showed me the way of the moose.

I stayed in homes, schools, or community halls. I never quite knew where I would be staying (like hitchhiking around Europe) until I landed on the gravel strip and was greeted by either the city or tribal administrator. And often, due to the weather, I did not know how long my stay would be. The longer my stays were, the more I could get into the subsistence rhythm of the villagers. Who needed help mending a net? A few times, I got to watch them skin a seal and help with the cooking.

I wrote in my journal, "The people know how to treat a stranger…make them feel at home then leave them be. I've been bathing in the warmth of people's generosity and the serenity of solitude."

I also wrote about the struggle to sell another federal program to Native Americans. The purpose of my trip to Togiak was to meet with village elder

Dan Nanalook and seek his permission to have a community meeting. His son David served on the Board of Directors of BBNA. David was also the village administrator that met my plane.

He drove me around the village. He wanted to show off the new school with the attached community hall. Then to his father's place overlooking the river. David proudly introduced me and I felt so honored to meet this revered elder, still going strong at eighty. He understood some English. I started to explain why I was there in the village when Dan, silhouetted by a kitchen window, interrupted me.

"You say this law can protect resources. It's the same with what they say about the wildlife refuge. Protect the land. Right?"

I nodded in agreement.

"What then about allowing bulldozing near an historic site?" Dan pushed. "Do you know about this?"

Stunned and nervous, I replied, "No. I don't work for Fish and Wildlife."

Dan harrumphed and waved me off. He then motioned for me to sit in the designated chair for visitors. Now on a roll, he explained how when the U.S. Fish and Wildlife first came to Bristol Bay, a long time ago, they were told it would be good for the animals.

"They come in here and say do no harm. Now they allow mineral leasing." Rising to his feet, Dan asked, "What good are white man's laws when they break them with their own law?

His words pierced me. I realized then that I was asking them to be part of "the system," which brought all the problems in the first place. Even though I had the best of their intentions in mind, I was still asking them to trust the system of the white man's laws. As far as Dan was concerned, all he needed was native law from the elders.

At the time of this six-month assignment with BBNA, the federal government was proposing to lease offshore oil and gas rights in the North

Aleutian Basin. This directly conflicted with subsistence and commercial fishing. As home to the world's largest sockeye fishery, commercial fishing in Bristol Bay was (and still is) big business and often accounts for half (in value) of Alaska's annual harvest. Fishing made up the economic lifeblood while subsistence fishing provided the cultural foundation. The proposed offshore area to be leased for oil and gas interests went right through the heart of the fishery.

Without the local control tool offered by the ACMP, the Bristol Bay region would not have a say in this federal leasing decision or in other development projects slated for this resource-rich region. They would lose a valuable voice for protecting subsistence and commercial fishing. It was a voice they never had before.

"Well sometimes the oil companies only know white man's law; only listen to white man's law. I'm offering a white man's law that could possibly do the same thing, only it would be to make them listen to your voice."

Dan seemed intrigued.

"You should think about it as back-up." I shrugged. "In case they don't honor native law."

In Yup'ik, Dan spoke to his son and other relatives that had wandered in. He rattled on in a most expressive manner, pounding the table for emphasis at key points. While I still have no clue what Dan was actually saying, I admired the passion with which he spoke. He seemed to be a genuine protector of his people's interest.

In reflecting back now, Dan Nanalook's passion reminds me of an earth elder familiar to many in the Pacific Northwest. Here are some words written by Chief Seattle in 1852 as a response to a U.S. Government inquiry about buying tribal lands.

> *The President in Washington sends word that he wishes to buy our land. But how can you buy or sell the sky? The land? The idea is strange to us. If we do not own the freshness of the air and the sparkle of the water,*

how can you buy them?

Every part of the earth is sacred to my people. Every shining pine needle, every sandy shore, every mist in the dark woods, every meadow, every humming insect. All are holy in the memory and experience of my people.

The shining water that moves in the streams and rivers is not just water, but the blood of our ancestors. If we sell you our land, you must remember that it is sacred. Each glossy reflection in the clear waters of the lakes tells of events and memories in the life of my people. The water's murmur is the voice of my father's father.

The rivers are our brothers. They quench our thirst. They carry our canoes and feed our children. So you must give the rivers the kindness that you would give any brother.

If we sell you our land, remember that the air is precious to us, that the air shares its spirit with all the life that it supports. The wind that gave our grandfather his first breath also received his last sigh. The wind also gives our children the spirit of life. So if we sell our land, you must keep it apart and sacred, as a place where man can go to taste the wind that is sweetened by the meadow flowers.

Will you teach your children what we have taught our children? That the earth is our mother? What befalls the earth befalls all the sons of the earth.

This we know: the earth does not belong to man, man belongs to the earth. All things are connected like the blood that unites us all. Man did not weave the web of life, he is merely a strand in it. Whatever he does to the web, he does to himself.[25]

Chief Seattle was right in that man did not weave the web of life. However, man does weave the web of political and social life. This was the conundrum brought on by my encounter with Dan. How could I work the complicated web of resource politics to protect the web of life in Bristol Bay?

I had joined the system and was representing it to a highly regarded elder who rightly did not believe in the white man's laws. My heart said, "More power to you," but my head said, "It won't work." They needed to have the tool of coastal management if they were to have any voice at all in offshore oil and gas leasing.

If I was to represent the coastal management system as proposed by the State of Alaska, I needed to be as honest and forthright as possible but I needed to do this without overselling the control offered by the ACMP and its federal counterpart, Coastal Zone Management. The requirement for federal agencies to be consistent with state approved plans was one thing, but extending that requirement down to local plans favoring subsistence resources was entirely unplowed ground. I decided I just had to be honest about both the potential and the limitations.

Obviously, my encounter with Dan Nanalook stirred up a lot of thought. Despite the initial unsettling, a game plan was starting to come together. After leaving Dan's home, I had David, his son, drop me off by the waterfront. Still disturbed from Dan's challenge, I needed to clear my head with a good walk.

By way of a flyer, I found my way into this community hall to laugh with six Yup'ik women about oosiks. This was so what the doctor ordered.

Two young men arrived, bringing round drums of stretched walrus skin over an 18-inch wooden ring. Yup'ik drums are struck with sharp, long sticks, producing a loud, crisp sound.

The grandmas handed out sewn grass rings with woven feathers; one fan for each hand. One dance sequence depicted the motion of fishing; another involved harpooning a seal; and a third was about scouting for caribou. The movements were sharp, decisive motions of the hands and head. The few words of each song were chanted over and over again, harmonizing with the pulsing drumbeat. The feet were to be stationary. They explained that Eskimo dancing is not for letting go but for becoming rooted and celebrating the subsistence rhythm of life. The feet were to remain rooted in the Earth.

I joined the dancing and kept my feet movements to a minimum. I learned that as long as I stayed rooted to the Earth I would do right by these elders. As long as one keeps hearing the voices of Earth elders, like Chief Seattle, one can find a path among the political thickets.

Renewed by the dancing, I went on to plastering coastal management posters all around the Bristol Bay region. I collected fourteen resolutions from Village Councils and one city resolution from the City of Dillingham. The question of establishing the Bristol Bay Coastal Resource Area passed by a sizeable margin on Oct. 6, 1980.

The plan did, without any involvement from me, include important provisions to protect subsistence resources and to raise questions about placing a lease sale in the middle of the world's largest sockeye fishery.

The North Aleutian Lease sale was last actively considered for oil and gas leasing in 2011. Then in December 2014, President Obama signed a presidential memorandum that will indefinitely protect Bristol Bay.[26] In declaring Bristol Bay off limits to offshore leasing, the president noted, "Bristol Bay has supported Native Americans in the Alaska region for centuries. It supports about $2 billion in the commercial fishing industry, supplies America with forty percent of its wild-caught seafood. It is a beautiful natural wonder, and it's something that's too precious for us to just be putting out to the highest bidder."

Sounds like Obama could be hearing Chief Seattle after all. All I know for sure is that when President Obama visited Bristol Bay in September of 2015 and danced with Yup'ik children at a local elementary school, he did not move his feet. Google "Obama dances with Alaskan children" and you will see that he remained rooted to the Earth.

Chapter 6 — Mountains, Missions, and Milestones

Figure 6: Kate took this photo after getting back down to 14,000 feet, post storm. It is unusual to see such a well-defined cloud cap over the 20,320 foot summit of Denali. A cloud cap is formed by the cooling and condensation of moist air forced up and over the peak and is lenticularly shaped by upper level winds.

There is no other mountain in the world like Denali. Although much lower in overall elevation from base to peak, it is more massive than the great peaks of the Himalaya and Karakoram. Mt. Everest has a vertical rise of 12,000 feet from the Tibetan plateau to its peak at 29,029 feet. Denali rises 18,000 feet above the Alaskan tundra to its peak at 20,320 feet. Mt. Everest is approximately 4-1/2° of latitude north of the Tropic of Cancer, but Denali is 3-1/2° south of the Arctic Circle. This far northern position not only makes Denali bitterly cold, but its proximity to the Aleutian Islands, the birthplace of winds, gives it severe weather.

Knowing the magnitude of the Denali challenge, I set about early to plant the seed of climbing with my circle of outdoor adventure friends. Several of them liked the idea of climbing Denali as a collective, turning-thirty event. But when it came time to commit and train, I found I was the only one still wanting to do so. As such, I looked into guiding options.

I joined eight men from across the Lower Forty-eight for a climb up the West Buttress, the standard route up Denali, or, as I prefer to call it, "the Washburn Route," named after its discoverer, Bradford Washburn. Slogging up 15.5 miles and 13,500 feet of elevation, the total time required to summit and return is from fourteen to thirty days long. Our team was aiming to take twenty to twenty-five days for the climb.

By the time the ski plane landed at 7,000 feet, I was in the best shape of my life. At the time of my training I was living in Ketchikan, a city of 13,000 in southern Southeast Alaska. Ketchikan, a long narrow city up against the mountains, is known for its many stair streets. I adopted one of these stair streets as my personal training area. I started with running up and down a flight of seventy-seven stairs ten times with nothing on my back, and slowly worked up to running the steps ten times in a row with a fifty-pound pack on my back. Although I could no longer see the mountain, the majesty of Denali held strong. Mentally, I was in the same place: Denali means *Great One* in the language of Athabaskan Indians.

The international scene at base camp was most entertaining. The French, German, Japanese, and Canadian climbers were, like me, all awaiting team members unable to make it in before the weather window closed. We passed time through storytelling. Inevitably, the stories evolved into a game of match-this-feat. Listening to their tales I felt that what I was doing was of little consequence, except, I supposed, to me and those with whom I shared my life. I realized then that consequences, like beauty, are always relative.

The climb was soon underway. As we eleven (including myself and the two guides) Americans headed up the Kahiltna Glacier, so too did ten members of the German Alpine Club. It was soon evident by their pace that we were going to have them as neighbors throughout our ascent. At first I was a bit dismayed not to have the ultimate Alaskan mountain more to myself, but I eventually warmed up to the idea. With a hundred-plus climbers on the mountain at once, it was unlikely we would have it to ourselves anyway.

We spent the next nine days setting up three camps, constantly moving toward our fourth camp at 14,200 feet. On an expedition climb, two trips are required to move gear, food, and reserves from one camp to the next. Essentially, you climb the mountain twice! I carried fifty pounds on my back and pulled fifty pounds on the sled. The days alternated between blistering hot, clear, windless days to intermittent cloud cover and wind—perfect conditions to make time. But at night temperatures dropped to minus twenty.

Eventually we reached the more established camp at 14,200 feet. Arriving there brought on a mixture of summit fever, high altitude weariness, and a little bit of spiritual giddiness. When I walked around camp with my tape player, I made a point of playing classical music to match the power of the setting. Peering down on valley glaciers and a myriad of rugged peaks, I was literally and figuratively on top of the world. Feeling deep contentment, I basked in Denali's glory.

The camp at 14,200 is a time when each climber needs to personally check the self-confidence gained from climbing the lower half of the mountain with the humility and courage for the formidable upper half. This high altitude camp represents the upper limit of one's ability to acclimate to the thin atmosphere at high altitude. From here on one's own reserves and mental toughness must carry one onwards and upwards.

Taking two days to acclimate at 14,200 feet helps avoid altitude sickness, an ailment that can strike any climber regardless of physical fitness. Altitude sickness is also known as mountain sickness and it ranges from a mild headache and weariness to a life-threatening buildup of fluid in the lungs or brain. Acute altitude sickness is the mildest and most common form. A more serious form of altitude sickness is high altitude pulmonary edema (HAPE). This illness occurs when fluid builds up within the lungs, a condition that can make breathing extremely difficult. Usually this happens after the second night spent at a high altitude,

I found that hanging out at 14,200 feet was like waiting for the mountain spirits to select who would have the privilege of making the final ascent. Luckily for us, we had no signs of altitude sickness in our entire climbing party. We all passed the altitude adjustment test and the first notions of summit fever began to percolate in conversations around the camp.

During our climb to 17,200 feet, our highest and last camp, the climbers with the German Alpine Club were a few hours ahead. The wisps of cirrus ponytail clouds signaled a dramatic change in weather: high winds were coming. As such, we were surprised to learn that the German Alpine club had pitched their tents and gone on for a summit attempt. Unlike the Germans, our guide, Nick Parker, had recognized the power and danger of the storm about to crash down on us. We dug our tents in deep and built wind walls out of the hard encrusted snow. Our tents fortified with snow anchors, ice axes, and snow walls, we set in to wait for the real storm to come.

At about midnight, the gusts pummeling our tents rose suddenly in intensity. Nick estimated the blasts to be in the 80 to 100 mph range. He barked out commands to wear or strap on every piece of clothing and equipment. In case of complete tent failure, we needed to be ready for quick evacuation. Within the next hour, the wind shredded our tent fly. The wind ground the snow to the consistency of talcum powder. Before long, without the fly, snow began creeping in through the seams. It was snowing inside our tents.

Once the winds hit, there was no longer any talk of the summit that was only an eight hour climb away. Not even a whisper. Nick went out to check the snow anchors that held our tent in place. He returned with the news that the German tents were gone, blown off the mountain. They still had not returned.

During the early hours of the storm, I worried about the Germans. But as the mania of the wind jostled me relentlessly about, my worries changed. *Forget the Germans. How in the hell do we get out alive?* I braced my body against the side of our tent, knowing that the only thing keeping the tent intact was our collective counter-push. My head braced against the tent felt like a basketball in a wild scramble for a loose ball.

I had nowhere to go but deep inside myself. I sought my core strength, trying to hold back the sense of doom and panic that slithered at my edges. I focused on constantly repeating this survival mantra: *I will make it through this storm. I have too much life force to die here. I will have children. The seeds for great accomplishments in my field lie within me. I will be a bridge of harmony for man and nature. I will live through this high-velocity hell. I have too much purpose of life.* At five in the morning, the Germans stumbled into camp, exhausted, disoriented, and some with frostbite and injuries from a fall into a crevasse.

The instant we stepped outside our tent to help, the wind snatched it away. One by one and within minutes, the howling, vicious wind took away each tent. Absent tents, we turned to the nearby snow cave currently inhabited by five Canadian climbers. They were aware of our situation and already in rapid digging mode. Fueled by adrenaline we turned a five-person cave into a twenty-three-person snow cave in relatively short order. "Ice condo cram" is what I dubbed our new home.

Our survival now depended on cooperation. We needed a system of sharing time in and out of the cave. The strongest worked to keep the entrance clear, others kept trying to shovel out more space from inside, and the injured rested. My job was to keep the hot drinks flowing and an eye on those the entrance most exposed, making sure they came in out of the wind every fifteen minutes. Nick, our guide, monitored the radios and tended to the injured.

The second night shoulder to shoulder with twenty-two other climbers brought on a new challenge for me: claustrophobia. It was a frightening, restless night. *I will have children. The seeds for great accomplishments in my field lie within me. I will be a bridge for harmony of man and nature.*

The next morning, the winds died down to about 50 mph. In this lull from the ferocity of the storm we hastened to escort the four injured Germans down to 14,200 feet, where they could be evacuated by high-altitude helicopters. No summit for us, but survival for all.

Figure 7: Kate on Pika Glacier, southern slopes below Denali, 2013. This was Kate's first time back in the Denali area since her climb in 1982. Photo courtesy of Nora Smith, Painted Post, N.Y. 2013

Upon my return home, my husband asked the question I was ready for, "Do you want to start a—"

I cut him off. "Yes, I want children now." There was no longer any hesitation about my goals in life. I went after them with renewed vigor. Bill could have asked me if I wanted five children and I would have said yes.

Two weeks later I learned that I was pregnant. Tracing back the likely day, I concluded that we conceived our daughter on my first day back from the mountain. Denali was consequential to me in a way I never would have imagined. Three months later, the consequences began to show. Soon came the stares and comments of surprise: "You're pregnant? Didn't you just come back from your climb?" To which I patted my stomach and responded, "The Mountain made me do it."

Chapter 7 — Here's to Strong Women in Sustainability

Figure 8: Kate with her lady friends kayaking in Berners Bay, 2016. Lion's Head Mountain looms on the horizon and yes, the water is very cold. Photo courtesy of Deb Loveid, Juneau, AK.

As a woman who has twice experienced the magic miracle of giving birth, I know there is an innate connection to the desire to nurture and protect. What I did not expect to encounter was that this innate desire to be protective could extend to looking after my guys in a commercial fishing fleet.

In 1989 I was hired as Executive Director for Southeast Alaska Seiners Association (SEAS), a major fishing trade organization. In this position, I was also on the board of the United Fishermen of Alaska (UFA), the statewide umbrella group for 23 regional fishing organizations. There I joined other women directors of regional fishing organizations who taught me the vernacular of "my guys"—as in my guys need to fish two on and two off (a reference to days fishing). Before long, I found myself advocating for my guys as in "that would help my guys too." Then the more I got to know the back stories of all those cute skippers, the more they did become my guys.

I came on board at the time that the management of the Tongass National Forest, the nation's largest national forest, became a hot issue before the U.S. Senate Lands Committee. Before I get too far into describing the conservation issue let me explain what a Southeast Seiner is.

A Southeast Seiner is a 58-foot fishing boat that deploys a seine net to corral and purse up thousands of pink salmon at a time. A seine is a fishing net that hangs vertically in the water with its bottom edge held down by weights and its top edge buoyed by floats. Seine nets can be deployed from the shore as a beach seine, or from a boat. Seine boats are designed to catch thousands of fish in one haul. In the 1990s, the annual harvest of pink salmon was about 49 million fish according to the Alaska Department of Fish and Game.

In 1990, the Senate was fiercely debating the Tongass Timber Reform Act (TTRA). This legislation set about to revoke artificially high timber targets and to require the Forest Service to cease the overharvest of rare, high value forest stands, bringing the Tongass National Forest closer to multiple use management, a management mandate required on all national forests. The TTRA also aimed to protect over one million acres of watersheds, and to require minimum 100-foot no-cut buffer strips on all salmon and resident fish streams. My assignment was to secure the much needed fish protections in an era of large-scale clearcutting.

Fortunately for me, SEAS was also a key organization within United Fishermen of Alaska (UFA). The first step was to engage UFA in the Tongass issue. Once we did, my voice grew from representing 500 fishermen in Southeast Alaska to representing 18,000 fishermen, the nation's largest commercial fishing organization.

UFA had a well-positioned D.C. lobbyist, Deming Cole. With an invitation for UFA to testify before the U.S. Senate Lands Committee, I was soon on my way to Washington. But first came Mom duty. Back in 1990, I was the mother of two young children about to set out on a long-scheduled family trip to Hawaii with grandma and grandpa. Squeezing in a trip to D.C. meant I would go direct from Hawaii and would arrive only the evening before. I'd have no prep time. I would be totally relying on Deming as UFA's lobbyist to have my oral testimony prepared.

Deming and I were unable to connect prior to my arrival in D.C. We planned to meet the next morning. I thought I'd have several hours to practice my testimony. The next morning, however, Deming and I discovered that we had crossed wires. "There is no prepared oral testimony," he told me. "I was leaving that up to you. No worries. We're going to my office. You have three hours." Instead of acting out the fit that was raging in my mind, I got Deming to step up and help me with secretarial tasks.

I had to get the importance of protecting stream habitats down to five minutes. I had to deliver the science and stress the economic lifeblood that salmon was to all the communities of Southeast. The only chance to practice my testimony ended up being with the cab driver on my way to the Russell Senate Building. Fortunately, he liked it. One down.

By the time we arrived, the wood-paneled hearing room was jam packed, standing room only, TV cameras and lights off to one side. The witness table with one empty seat seemed far away. My heart was *absolutely* racing. To calm down, I switched my focus to music, a driving rhythm in my head. I tuned into the fourth movement of Beethoven's 5th. I wove my way to the witness table. Return to playing Beethoven's 5th. I started talking to myself. *No surprise, I'm the only woman. Don't let go of the music until you're ready*. I took my seat. *Breathe. Music off. Engage. Breathe again.*

The six senators of the Lands Committee entered and took their seats behind a podium looming large above me. My pulse rose. I looked around and realized who was sitting two chairs away from me, my hometown adversary, Don Finney, representing the Alaska Loggers Association. *Oh, sweet!* I thought, *I speak fourth, after him. I bat clean-up.* Suddenly, I felt calm, ready to get on with business.

I knew in advance that the hard part would be to get through the questions of the committee members without looking like a fool. If I could score a few points, all the better. Senator Frank Murkowski from Alaska would seek to poke holes in my testimony wherever he could. He threw a softball question to the Alaska Loggers Association representative. Satisfied, he then attempted to discredit the science behind the 100-foot stream buffers. Before answering, I went back and answered the lob question thrown to Don and spun it all the way around. I didn't get riled. The science supporting buffers was on my side and strong. Another question. No problem.

In their closing remarks, a Democrat from New England said he liked my terminology of "eco-nomics." He asked his staff to make note of that term. Hearing this comment left me beaming.

According to Deming, I held my own against Senator Murkowski's attack on the fisheries science. Deming and I shared a high-five. "Today is a day," said Deming, "we did good for salmon." It was no longer just good for my guys. The maternal desire to nurture and protect now extended to salmon. I did my part. I wanted to go out and celebrate but Deming was booked for the night.

Thirty-five years later, I celebrate every time I cross a salmon stream in the Tongass National Forest. After reading award winning journalist Naomi Klein's chapter, "The Right to Regenerate" in her book, *This Changes Everything,* I now think of myself as having delivered for the right of salmon to regenerate. That's what I celebrate with every wooded stream crossing.

As Ms. Klein, among other activists, rightly expresses, "All of life has the right to renew, regenerate and heal itself."[27] According to Klein this message is penetrating into a movement, "What is emerging, in fact, is a new kind of reproductive rights movement, one fighting not only for the reproductive rights of women, but for the reproductive rights of the planet as a whole – for the decapitated mountains, the drowned valleys, the clear cut forests, the poisoned rivers, the cancer villages."

Indigenous tribes in Bolivia and Ecuador have even enshrined the *Right of Mother Earth* into law, creating new legal tools to assert the right of ecosystems not only to exist, but also to regenerate."[28] This *Right to Regenerate* is also enshrined in the *People's Agreement* of the April 2010 UN Climate Conference in Bolivia.[29] The agreement asserts that the earth has *the right to regenerate its bio-capacity and to continue its vital cycles and process free of human alteration.*

While this principle of being "free of human alteration" is not possible for thousands of fish and wildlife populations, it is a noble principle to assign to the earth, and a noble endeavor to codify if possible. The City of Pittsburgh, Pennsylvania is the only American city I know of that has come close to doing this. In 2010, the Pittsburgh City Council passed a law banning all natural gas extractions and stating that nature has "inalienable and fundamental rights to exist and flourish" in the city.[30] In essence we now have a major U.S. City acknowledging the principle behind the U.N.'s People's Agreement. Until I read Ms. Klein's book, I had no idea that the *right to regenerate* movement even existed, let alone is becoming enshrined on a global scale by the United Nations.

Of any group of species whose right to regenerate should be respected, it is salmon. Salmon, after making an epic 2,000 mile journey from the gyre in the Gulf of Alaska to the open ocean, and back past fishing nets, fish hooks, bear claws, and eagle talons to finally reach their natal stream, deserve their right to regenerate, their right to spawn before they die. And when salmon regenerate here in Southeast Alaska, we all do. It's enshrined not only in law, but in the hearts of our community. In the large expanse of Alaska, I am fortunate to experience a world relatively free of human alteration. But with this privilege comes a sense of responsibility to do more, to help beyond the salmon.

Regardless of where we live, the profound connection between women and the environment is creating a new right for mother Earth. Furthermore, research shows that women are also more prone to give back politically and in their decision-making. Kira Gould and Lance Hosey, authors of *Women in Green*, suggest that women show more support and do more for the environment. To make the case, they cite these survey results:

- *Polls consistently show that women are up to 15 percent more likely than men to rate the environment a high priority.*

- *In political elections, women comprise up to two-thirds of voters who cast their ballots around environmental issues.*

- *Women are more likely than men to volunteer for and give money to environmental causes, especially those related to health and safety within their own communities.*

- *Women report both more support for environmental activists and more concern that government isn't doing enough to protect the environment.*

- *More women than men support increased government spending for the environment, while more men favor spending cuts.*

- *Women tend to be less lenient toward business when it comes to environmental regulation.*[31]

These stats show that regardless of the motivation, whether it be maternal or not, when you tap into women, you're more apt to connect into more good for the environment. With more women entering the workforce, this ability to connect will only improve.

According to Hanna Rosin, a journalist and contributor to *The Atlantic*, the global economy now favors women. In a TED talk, entitled, "New Data on the Rise of Women," Ms. Rosin reports that "Women, for the first time this year became the majority of the American workforce.[32] And they're starting to dominate lots of professions—doctors, lawyers, bankers, accountants. Over 40 percent of managers are women. So the global economy is becoming a place where women are more successful than men, believe it or not, and these economic changes are starting to rapidly affect our culture."

It is no accident that the number of women in the workforce is rising. The global economy has shifted. We used to have a manufacturing economy centered on producing goods and products; now we have a service and information economy. As Ms. Rosin points out, "These new economies require very different skills and as it happens, women have been much better at acquiring the new sets of skills than men have been. This new economy is pretty indifferent to size and strength (valued in a manufacturing economy). Now you need an ability to sit still and focus, to communicate openly, to be able to listen to people and to operate in a workplace that is much more fluid than it used to be, and those are things that women do extremely well."

Along with the greater presence of women comes a different set of personal values; a willingness to consider how an action will affect other people instead of simply asking, "What's in it for my company?" Community, a long-term perspective—these are values that women leaders tend to bring into the working world.

In business, women are actively creating a better business model. According to Sally Helgesen, the author of *The Female Advantage: Women's Way of Leadership*, "Business models based on traits often identified as feminine, can ensure the on-going viability or sustainability of not only business but also any community—as well as all of humanity."[33]

Contrast this perspective with the more conventional perspective of Wall Street businessmen, which is best summed up by Nobel prize-winning economist Milton Friedman, "The business of business is business."[34] Without a sense of community, of social responsibility, it misses half the world. The business of business needs to be community.

Although many male CEO's embrace social responsibility and incorporate sustainability models into business, new research show that these models of business do better with gender-inclusive leadership. Researchers from the Harvard Business School and a non-profit, Catalyst, (dedicated to creating diverse workplaces) found that more women leaders in companies is correlated with higher levels of Corporate Social Responsibility (CSR).[35] Corporate social responsibility means that company committed to CSR acts as a good corporate citizen, expanding the definition of success beyond impact to the bottom-line to also consider the organization's impact, both positive and negative on the world.

> *Companies are realizing that advancing more women to senior leadership roles has many benefits, including increased financial performance and sustainability,"* said Anabel Pérez, Senior Vice President, Development, Catalyst. *"As this study shows, inclusive leadership has a positive influence on the quantity and quality of an organization's CSR initiatives. When business leadership includes women, society wins.*[36]

Bottom-line: gender-inclusive leadership is good for business and society. Can this be synergy – eco-nomic principle number eight – at work here? The study by Harvard Business School and Catalyst concludes that "companies with both women and men leaders in the boardroom and at the executive table are poised to achieve sustainable big wins for the company and society".

Male leaders that know the value of gender-inclusive leadership score big. One champion of promoting women in sustainability is Ray Anderson, once dubbed by Forbes Magazine as the "greenest CEO in America." Here is what he had to say about how the future of sustainability will depend on women:

A new day dawning will build on the ascendancy of women in business, the professions, government, and education. This is one of the most encouraging of all trends, as women bring their right-brained, nurturing nature to bear on the seemingly intractable challenges created by left-brained men and their pre-occupation with bottom lines and other "practical" considerations. After all, it is the practical and pragmatic that got us into this mess. Surely, a different kind of thinking is needed to get us out.[37]

Mr. Anderson said this about the ascendancy of women in 2005, five years before women became the majority in the American workforce and filled 40 percent of the managerial positions. The new day Mr. Anderson saw dawning was eight years before a report by the Harvard Business Review, "The Global Rise of Female Entrepreneurs", which concluded that women entrepreneurship is a *fundamental economic force that's reshaping the world* because women are re-defining the business equation to be more inclusive while increasing the bottom-line at the same time.[38]

Ms. Jackie VanderBrug, the author of the report, writes, "The recent Global Entrepreneurship Monitor found 126 million women starting or running businesses and 98 million operating established businesses for a total of 224 million women impacting the global economy."[39] These are huge numbers. Think about the fact that women entrepreneurs now represent 37 percent of enterprises globally. Now imagine all of these women using their right-brain to find and implement new ways of practicing community-based business models. It works around the world.

Marge Piercy, a poet, writes: "A strong woman is strong in words, in action, in connection, in feeling; she is not strong as a stone but as a wolf suckling her young. Strength is not in her, but she enacts it as the wind fills a sail."[40]

Around the world, strong woman are filling their sails, enacting change on many levels, making a difference in their communities and to their families. One of the first environmental movements inspired by women was the women tree-huggers in India. They were called the *Chipko movement*. "Its name comes from a Hindi word meaning to 'to stick' as in glue. The movement was an act of defiance against clearcutting a community watershed. Their slogan is: *Ecology is permanent economy.*"[41]

If you recall my concept of eco-nomics, you'll note the similarity. I spoke about eco-nomics as a sound economy and a sound environment going hand-in-hand; stability in the ecosystem system promotes stability in the economic system. However, I like the Chipko women's take on eco-nomics. *Ecology is permanent economy.* That's a better way to explain it.

Chapter 8 — Turbulent Waters

Figure 9: Kate rowing on the Tatshenshini River, British Columbia, 2010. Photo courtesy of Bill Hanson, Juneau, AK.

My most harrowing kayaking experience came where I least expected it
. . . in the inside, protected waters of Glacier Bay National Park. Shortly after
my move to Juneau from Ketchikan in the summer of 1992, I had an
opportunity to get an affordable kayak trip to Glacier Bay. I could take time
off from working for SEAS. In assessing the home front, my two children,
Erin and Rion, seemed to be adjusting to a new elementary school. We even
had cousins down the street that could help my husband with kid duties.

The Glacier Bay trip offer came from Alaska Discovery, a guided tour
operation. They asked if I, along with three friends, would be interested in
transporting a couple of two-person kayaks from the mouth of Muir Inlet
to McBride Glacier near the head of the Inlet. We would have a free trip to
Glacier Bay and the opportunity to help Alaska Discovery by moving the
kayaks from the end of one of their trips to the beginning point of a National
Geographic charter trip. They would drop us off by boat, and we would fly
out on the airplane that brought in the National Geographic crew.

I asked about the distance. We had four days to kayak a little over thirty
miles. I noted this timeframe didn't allow for bad weather. Where was the
slack? I was told we need not worry about weather at this time of year. It was
a doable distance. I just needed to find crew.

My outdoor adventure girlfriends from Ketchikan, Rosie and Christine,
came right to mind. Ketchikan is a town of 13,000 residents about 230 miles
south of Juneau. Rosie was all in straight away but Christine had to sort
through a potential work conflict. Rosie and I cast our nets wide for a fourth
woman strong enough to kayak at least thirty miles in four days.

We discovered that Christine could only go on the trip if she met up
with us the next morning. There was an 8 a.m. tour boat leaving Bartlett
Cove that dropped off kayakers. This would reduce our paddle time by at
least a half-day. Rosie found a nurse, Nancy, who had already scheduled
time off and best of all, she was an experienced kayaker.

Boarding the Alaska Discovery boat in Bartlett Cove, there was a nip in
the air. It felt breezier than I would have expected in this protected cove.

Once out in Glacier Bay, the rain-laden clouds settled in. But my mind didn't settle. It was restless like the wind off the port bow. The four-foot waves stacked up as we entered Muir Inlet. Our drop off point was somewhere out there by Sebree Island, the gateway to Muir Inlet.

Once we were behind the wind break of Sebree Island, I switched off the worry channel and found calmness in the sight of an accessible beach with two double-kayaks tied off in the tall beach grass. We quickly off-loaded.

One of my favorite moments in the wilderness is when the boat or airplane engine can no longer be heard, leaving only the silence of your soul to fill the space. The cry of an eagle provides the perfect accent. A loud "caw" of a raven has me bowing to the moment of being enveloped by wilderness.

Rosie, Nancy, and I quickly set up camp and managed to eat before the rain started in earnest.

"We'll need to get up early to pick up Christine at Tlingit Point. It's a good mile away around the back of the Island," I said.

"Puts us back in that mixing bowl?" replied Rosie. "No thanks."

"I agree. Hey what if we call the tour boat skipper by radio? See if he can drop Christine off on this side of the island?"

"Do you think our handheld would reach him?" Rosie pondered.

"Not until he's pretty close. Worth a try. This is a much easier place to disembark."

The next morning white-capped waves testified to the whipping wind out in the middle of Muir Inlet. We were convinced that the skipper would need to drop Christine on the lee side of Sebree Island for our safety, if not for the ease of his own landing. I raised him on the radio.

"That's a negative," the skipper told us. "I can't deviate from the Park Service-approved itinerary."

"Not even under these conditions?"

"You want to get your friend; she'll be at Tlingit Point. If not she can run back to Bartlett Cove with me."

I considered the options. We'd have to paddle out of calm waters and into challenging conditions. Was it doable?

"Can I talk to her?" I asked the skipper.

"Christine, what do you want to do?"

"Land. Be with you guys."

"We'll leave now for you. It will take at least thirty minutes."

In launching the kayaks, Rosie offered to paddle the double with the empty bow seat. She was certainly strong enough. However, we needed to add ballast to her bow to keep if from bouncing around in the waves. Everything needed to be snug. We cinched our spray skirts as tightly as possible to prevent the waves that would be breaking over our bows from filling the kayaks with water. We needed to "wear" our kayaks as much as possible by bracing our knees against the sides. Hip control would be essential in paddling through the big water ahead.

As soon as we poked our heads out of the cove into the inlet, we were slammed by three-to-four-foot waves. Heads down, we stroked hard. With the backwash off the rocks, waves started coming from three directions. The waves head-on were building to six feet. Angling to avoid the building waves meant taking some of the backwash on our stern.

Surfing down a wave, Nancy in the stern of my boat let out a big "Yeehaw!" I heard Rosy scream the same.

"Yeehaw!" I joined in. "Yeehaw!" I felt my fear lessen. I relaxed the power grip I'd had on my paddle.

Well into the mixing bowl, Rosie's kayak began to go up and over the waves instead of slicing through the crests. Without enough ballast she had a hard time controlling the kayak, suddenly at risk of capsizing if a wave pushed her broadside. Nancy and I tried to stay in sight of Rosie at all times. No more Yeehaws. We had now come abreast of Tlingit Point. When we turned ninety degrees toward shore, we would momentarily be broadside to the waves. Our timing would be critical.

"On the next wave," I screamed to Rosie. She came back into view on the crest of an intervening wave. I saw her snap her hips and turn her bow just has the water rushed over it. She slipped down into the next trough. She signaled *I'm okay* with her paddle.

Nancy and I stroked with mad intent. We turned, ruddered hard, and back-paddled. A wave buried Nancy in the stern. She braced with her paddle. I gave three hard strokes and we were momentarily clear of danger.

Christine had spotted us. Abandoning her pile of blue dry bags on the rocky beach, she ran along the shore to the nearest landing spot. I was elated to see Christine, but even more excited to see the relief and smile on Rosie's face, radiating the joy of having made it through the mixing bowl and around the point.

The return journey around the point was much easier for Rosie with Christine in the bow, both because the kayak was properly balanced, and because of the power of a second paddler. Although the tide change had calmed the waters some, the waves were still four footers with foam on top. We needed all of our strength and concentration.

An hour later we arrived at our camp. Needing to travel at least five miles toward our destination we rapidly broke it down. This was day one of the trip. Days two and three were ten-hour days of paddling in non-stop cold rain. At times, the persistent rain flared up and came in bursts so intense that visibility became an issue. On any other trip this would have been an automatic bad weather day to sit out. Our land breaks were held to the minimum time necessary to empty our bladders.

By the third night of camping, my sleeping bag, along with Christine and Rosie's bags, was damp throughout. Nancy's bag was just fine. She had better gear and was snugly diving into a book that still had dry, turnable pages.

Christine, Rosie, and I decided to zip our bags together to maximize body heat. We couldn't afford another sleepless night of shivering. Body heat was now essential. Like teenage girls at a slumber party, we rubbed each other's hands, drank chocolate and schnapps, told stories, and giggled. With only a half-day of paddling left, arriving at McBride Glacier by noon was well in hand. We had made it through the big water, the relentless rain, and the cold of damp clothes. We had found the strength in our paddles to continue on. We had found the ability to laugh off discomfort and difficulties.

Our time at McBride Glacier seemed way too short. By comparison to all the time paddling, making and breaking down camp the last four days, four hours of non-work time in light rain seemed hardly fair. We barely had any chance to hang out at all, relax, and enjoy being in Glacier Bay National Park. The plane was due in by 10 a.m. the next morning. Although grateful for making it to McBride glacier in time for the plane, we felt a bit jilted by the circumstances surrounding the trip. First, no bad weather days allowed in the schedule. Then to lose half-a-day to getting Christine, to jeopardizing our safety when sent around the point. To only getting four hours of down time at McBride Glacier. Yep, life is not fair.

We all know this. We also know that my "life is not fair" moment could've been much worse. Yet, that doesn't mean we have to like these moments. Disappointment sucks, pure and simple. The challenge comes when disappointment moments start to stack up like a series of building waves, running hard at mid-tide. The question then becomes can you wait it out? Can you keep these moments or series of disappointments from becoming the whitecaps of bitterness?

Little did I know that the turbulent waters of Glacier Bay foreshadowed a time of professional set back when I least expected it. After recovering from the sogginess of the trip and all the parental duties I had left behind, I moved on to prepping for a possible big career move by jumping into the mixing bowl of gubernatorial politics.

I positioned myself well within the campaign and delivered the valued endorsement of UFA. Again, I got jilted by circumstances beyond my control. Since I don't want to make this part of my professional story, personal beyond me, details will be short in coming. In essence, when it came to gubernatorial politics, I did everything I was taught to do—write checks, bring along others, draft position papers—and I worked for the state's largest private sector employer, the commercial seafood industry. I had well-connected friends. At last, I was in the right place at the right time so I played all my political cards in an honest and straightforward manner.

Nothing came my way. I eventually got a classified position and ended up working for the poorly qualified man who got the appointment in my stead. Needing to steer him through fish politics left a sour taste in my mouth. I had to re-orient my thinking. Instead of engaging in big picture politics where I would only end up frustrated, I switched to planting seeds and taking baby steps toward helping individuals and small groups of fishermen. Yet, the taste of bitterness still lingered.

Needing an outlet for my vision, I turned to writing. I decided to write stories to go along with my nine principles of eco-nomics (see Appendix B). I wrote my first non-fiction book, *Eco-nomics and Eagles: A People's Guide to Economic Development AND the Environment*. Next came the humbling task of trying to find a publisher. After scores and scores of rejections, I ended up in 2002 self-publishing *Eco-nomics and Eagles* through Xlibris.

The book included an afterword; an afterword that reads more like journal writing. As you can see from this excerpt, I am still riding the unsettling waters of professional disappointments.

Since the [Denali] climb, I have had more than my share of career-path setbacks. In an effort to let go of some disappointments, I began to write this book. Then with publishing rejections coming in, the spiral down continued. Then we went on a family reunion camping trip to Denali National Park. As the mountain poked through the clouds at the Eielson Visitor Center, I found myself telling the children around me (nieces and nephews, too) my Denali story. I was able to point out how high I had gone. I was able to tell them of the inner belief that kept me strong up there in the jet stream storm.

For the first time in many years, I felt the pull of the second half of my survival chant. I will be a voice for harmony, for linkage. Why, if the first part, the pregnancy part, came true so soon after my climb, why not the second part? Why is the pivotal policy job eluding me? Could the mountain be wrong?

I continued to wonder if I would ever find the "yeehaw" again.

Chapter 9 — Soldiers of Sustainability

Figure 10: Small boat fishermen offloading their sardine catch with a flotilla of frigate birds ready to pick up any and all scraps. Some birds don't wait and just attack the tub of sardines. Puerto Lopez, Ecuador, 2005. Photo taken by Kate Troll.

I was barely functional when I arrived in Guayaquil, Ecuador, at two in the morning on my own to find my way around a new city. The taxi driver stopped in front of what appeared to be a downtown shopping mall, not a hotel. He insisted I follow him into the mall. As he marched me to a side elevator, where he left me to find my way alone, shoppers buzzed by in blurs of neon color bouncing off glass. *Was I in a teenager's shopping dream?* What seemed like an hour in a daze probably only took twenty minutes of wandering through stores and hallways to find my hotel room. Once there, I crashed.

When I awoke, I was shocked to see it was already twelve-thirty in the afternoon. I rushed downstairs to the lobby where I happily found my work associate: Ernesto, a forty-something single dad with an almost-grown daughter, a gentleman to all whom he met. He was my reliable, entertaining, easy-going partner.

I first met Ernesto at an NGO fisheries conference in Mar de Plata, Argentina, where he had come to hear me speak about the Marine Stewardship Council (MSC), an international eco-label for well managed fisheries. I was the regional fisheries manager for the America's office. Working for the MSC allowed me to find an interesting exit from direct engagement in Alaska's fish politics. I would no longer be working for the State of Alaska but would still interact with Alaska's fisheries as they sought certification of sustainability. I could still be connected but more on my own terms and I could go international in scope of global fisheries.

My region was all of North and South America. As a fisheries consultant working for several small fishing organizations, Ernesto was the professional colleague I had been secretly hoping to find—one who I could put on contract to work for me; one who would give me local credibility in setting up meetings. I quickly got Ernesto approved to assist me.

This was my third trip with Ernesto. As we travelled through Ecuador, I left him in charge of setting up our visits to coastal fishing communities. The main meeting of the day was to speak with a woman named Gabriella who was the head of the artisanal fisheries union. Artisanal fisheries are small-scale fisheries for subsistence or for local, small markets, and generally use traditional fishing techniques and small boats. They occur around the world (particularly in developing nations) and are vital to the livelihoods and food security of millions. Gabriella represented twenty-three of these artisanal fishing communities. She was leading her annual price discussions with local fishermen and processors. If I impressed her, we could accompany her on her annual trip along the western coast of Ecuador.

Off we went to meet up with Gabriella. When we walked into her wood-paneled office after waiting thirty minutes in a smoke-filled lobby, I thought I had entered a beauty salon. The air was thick with the smell of hairspray. Her office was made cramped and tight by her clutter. She motioned for us to sit in small aluminum chairs squeezed close to her desk. Gabriella primped her hair one last time before addressing us. At last we spoke, or more correctly, Ernesto spoke while I smiled graciously, as Gabriella clearly wanted to be admired. I also nodded my head every time Ernesto said "Claro."

At the end of the meeting I put my hand out to shake Gabriella's. She shook Ernesto's hand but waved me off, saying "Manana." Upon exiting Ernesto let me know that I passed her inspection, but just so-so. She would let me have ten minutes at each meeting to talk about the MSC and sustainable fishing practices.

At three a.m. the next morning, (*why always these early morning hours?*) Gabriella was to pick us up at the hotel to drive north to the town of Esmeraldas. Then we would head south to the main tuna port, Manta, where I had two meetings set up with international seafood companies.

Before going to bed that night in the Mall Hotel, I read about Esmeraldas. To my surprise, the *Lonely Planet Guidebook* said, "If you must go to Esmeraldas, perhaps the most dangerous city in all of Ecuador, here is what you need to know." I can't remember any guidebook every stating "If you *must* go there. . ." I immediately emailed my husband so that someone knew my whereabouts.

After turning off an incessant alarm, I joined Ernesto in the lobby. We waited. And waited. An hour later, Gabriella showed up in bright colors and holding onto her fisherman boyfriend. She offered no apology and quickly introduced us to Juan, the young muscle-bound man who would drive the car.

Ernesto and I squeezed into the back of a small, worn-out Honda civic. Gabriella, not letting go of Juan, slid CDs into the disk drive. Juan drove the stick shift rough and hard like a boat in shallow, wind-driven water. With zero leg room and a drive of seven hours this was going to be brutally uncomfortable. Gabriella insisted on cranking up the Mexican pop music. All this was too much for me and I became visibly annoyed.

Sharing my misery, Ernesto put his arm around me, shook my shoulders, and in a mock German accent yelled, "We are comrades-in-arms. We are soldiers of sustainability. We can do this."

I laughed and popped a melatonin sleeping pill into my mouth. After dozing off for two hours, Ernesto shook me awake. The car came to an abrupt stop by a roadside food stand.

"Breakfast!" he yelled.

"Can I just sleep? Please?"

"That's not an option. Gabriella insists we all eat now."

I shrugged Ernesto off and peered out the window, only to see Gabriella standing expectantly with her hands on her hips.

I turned to Ernesto. "Soldiers of sustainability?"

"Out this moment," he ordered.

I saluted him. "Soldier of sustainability reporting in."

Anticipating my next question, Ernesto pointed to the back of a nearby building. Heading to the restroom I nodded to Gabriella and said "Buenos días, señora."

Back in the car, we continued our miserable ride to Esmeralda where, according to Gabriella, we would meet with maybe twenty fishermen. As we pulled up to the main dock, the wretched yet familiar smell of fish offal greeted us. Gabriella quickly directed me into a break room off the offloading dock. During this meeting with a handful of fishermen, it soon became clear that my role was simply to smile and nod when I heard Gabriella refer to me as "la mujer internacional." Neither Ernesto nor I were given any opportunity to speak.

Although I had long before learned that empowering others empowers one on a community level, I still needed to engage directly with the fishermen to explain the MSC mission and opportunities. Ernesto understood this and managed to get Gabriella's assurance that we would have a chance to speak at the next meeting in the community of Lagarto.

In route to Lagarto, we turned off the potholed, paved road onto a dirt road that led to the tiny village of Rio Verde. Consulting the *Lonely Planet Guide*, I learned that Rio Verde was the site of the film *The Living Poor,* a Peace Corps chronicle. It was a film that aimed to educate volunteers about living in third world poverty. The wooden, partially falling huts along the river's edge evoked this title.

The village across the river from the fish plant was accessible only by a foot bridge large enough for two people or one person and a donkey to pass. We could go across but there would be no meeting; the fishermen here were not organized. They were too poor, their catches too low, for Gabriella to bother with. Still she needed to say hello to one fisherman; let him know she was here.

I followed her across the bridge, which seemed to barely cling to the riverbank. Beneath the walkway were a few dugout canoes loosely tied to a post. They had fragments of nylon gillnet in their sterns. As I approached the far side, more of the village came into view. In a calmer eddy stood a series of wooden shacks on stilts. At the far end stood a common outhouse in need of the high tide. Naked kids and kids in underwear played nearby.

Crossing back over the foot bridge an hour later, I felt relieved to be away from such stark poverty. I was profoundly bothered by the knowledge that our sustainable fisheries program was not designed to help the fishermen of Rio Verde, that it was out of their reach. I found myself needing to take solace in at least having a life with purpose.

Gabriella was again nowhere in sight. Ernesto and I were back at the car right on time. Two hours later, Gabriella showed up. With no explanation, we headed off at breakneck speed to our next union meeting in the town of Lagarto with Gabriella, shouting "No problema" each time the car lurched over the minefield of potholes.

This time she gave me fifteen minutes to explain my program through Ernesto's translation. Once out the door, she pulled out her cell phone. She motioned us toward the car. "How long?" I asked. She didn't answer.

Refusing to wait inside the car and sensing that this would be an invitation for Gabriella to once again keep us waiting and waiting, I paced back and forth. I could see that she noticed my irritation, yet she remained on the phone.

Exasperated, Ernesto, the good soldier of sustainability that he was, sprang into action. He headed back to the fishermen's hall where there was a telephone. A few minutes later, he waived me over to approve hiring a taxi from Lagarto to Puerto Lopez. It would take us seven hours at a cost about $150. One look back at Gabriella, who spun her conspicuous butt away from me: Time to change everything. From Puerto Lopez we could possibly catch a bus for my meetings in Manta.

"Change Everything Now" was the motto of the 2014 People's Climate March in New York City. This is a reference to changing the environment through re-defining capitalism and builds off the written work of Gus Speth who is a former administrator for the United Nation's Development Program. Mr. Speth is also the former Dean of the Yale School of Forestry and Environmental Sciences and the recipient of the Blue Planet Prize. In an *Orion Magazine* interview, journalist and author Jeff Goodell notes, "If

America can be said to have a distinguished elder statesman of environmental policy, Speth is it."[42] In other words if I am but a soldier of sustainability, Gus Speth is the general.

In a Jeff Goodell interview entitled "Change Everything Now," Gus Speth explains what a revised capitalist system would look like.

> *Maximizing the shareholder wealth is a very fundamental part of the motivational structure of the corporate sector. I think that needs to change fundamentally, so that corporations really are in the business of serving all the factors that help generate wealth—all of the stakeholders, in effect. One way to describe what has to happen would be a series of transformations. The first would be a transformation in the market. There would be a real revolution in pricing. Things that are environmentally destructive would be almost out of reach, prohibitively expensive.*[43]

Speth goes on to articulate ten more transformations in his book, *America the Possible, Manifesto for a New Economy*. He speaks of the possibility of a technical nirvana that would handle all the environmental waste and destruction.

> *Changes of the type that would bring on this technological nirvana are just too slow and too partial. They need to be combined with other things that basically slow the current up. And that means taking the priority off of growth. It means finding new laws for corporations—to change their incentive structure. It means us consumers becoming more interested in living more simply.*[44]

In a way, changing the incentive structure was what I was working on when working for the MSC, albeit in a small way. When I first started working for the MSC, their blue "well managed fisheries" label, was but one of a dozen eco-labels. Now there are 458 ecolabels in 195 countries.[45]

In an article entitled, "Questioning and Evolving the Eco-label" in *The Guardian* (March 2011), reporter Patrin Watanatada concludes that "ecolabels have proven the power of a good idea; an idea that combines sustainability standards-setting and branding, underpinned by the credibility of an independent body."[46] But how well has eco-labeling lived up to its ambitions? This is a very complex question, and results are difficult to quantify. A study done by the World Wildlife Fund (WWF) found "insufficient comparable and meaningful data available to determine performance on sustainability missions."[47] This is essentially the same conclusion reached by a previous study on a wide array of eco-labels done by the United Nations.

Without hard numbers, anecdotal evidence about changing behavior becomes more significant. As noted in *The Guardian* article, "Ultimately, ecolabels strive to accelerate sustainable behavior. Neither consumers nor producers can be expected to do the right thing unless they know what that is, and eco-labels are to be commended for focusing on this need— as are the global companies who are pushing to make effective use of them."[48]

One of those companies seeking to make good use of the MSC program was waiting for me in Manta, Ecuador. Yes, Ernesto and I had managed to find a bus from Puerto Lopez. We arrived in Manta just one hour ahead of a long-scheduled meeting with Atunec, an association representing eighty-seven tuna vessels. We discovered that the CEO of Atunec was extremely anxious to meet with me as he had been talking about funding an MSC Sustainability Workshop for the entire Ecuadorian tuna fishery. He was keen to invite tuna fishermen from California who were already engaged in MSC certification to share their knowledge. My meeting with him in Manta would set these wheels in motion.

Although pleased to make this connection with a large company seeking to move the needle on sustainable fisheries, I was still bothered by the MSC's inability to throw a market net large enough to capture the smaller, artisanal fisheries. After my experience in Ecuador, I joined my fisheries counterpart in the Asia-Australia office to put forth a proposal to help sustainable wannabe fisheries.

We suggested using the MSC network of buyers to benefit "sustainable fisheries in training" by agreeing to source a small proportion of their needed supply from these transitioning fisheries. Similarly, the MSC network of buyers could support fishermen who wanted to change management and operations to become more sustainable, but who could not get there without re-inforcement from the marketplace. These proposals used only the buyer network and did not require the use of the eco-label. Our proposal did not even make it to the MSC's Technical Committee.

Even though I may be disappointed in the MSC in this regard, I am nonetheless pleased that during my tenure with the MSC, I helped fourteen fisheries through the scientific assessment process. Now the MSC has 178 certified fisheries in all parts of the world.[49] My experience working with the MSC leads me to believe that eco-labels, though not perfect eco-reflections of sustainability, do foster change and awareness of sustainable practices.

The real force behind the MSC working at all was Unilever, the world's largest wholesaler of seafood. As the world's largest supplier of seafood they had the market share to make the MSC happen and the most to gain from sourcing seafood from sustainable supplies. Univeler was pioneering then and their CEO is now. In a TED talk (Jan. 2014) entitled "Profit's not always the point", Harish Manwani promotes the "Four G's of Growth":

> The model that at least I was brought up in and a lot of us do business is what I call the three G's of growth: growth that is consistent, growth that is competitive and growth that is profitable. And I'm afraid this is not going to be good enough and we have to move from this 3G model to a model of 4G. The fourth G is growth that is responsible. Companies cannot afford to be just innocent bystanders in what's happening around in society. They have to begin to play their role in terms of serving the communities which actually sustain them.[50]

Millennials, although not necessarily using the 4G language, have been pushing companies to become more socially responsible. According to an analysis done by Haas School of Business, University of California, Berkeley, "Over 80 percent of millennials will switch brands if the companies have no

Corporate Social Responsibility initiatives." This is further documented by a cause marketing agency, Cone Inc., which found that more than nine-in-ten millennials consider a company's social and environmental commitment when deciding where to shop and would switch brands to one associated with a cause (91% vs. 85% U.S. average)."[51]

Until we as "consumers become more interested in living simply," to quote Gus Speth again, where we shop will continue to be an important outlet and source of our eco-awareness. Consequently, the more the millennials switch brands for greater corporate responsibility, the better.

I'm glad to have played a part in fostering eco-awareness of seafood from well-managed fisheries. I'm glad to have made a friend in Ernesto. I'm glad we had the willingness to change everything when on our Ecuadoran adventure.

Chapter 10 — In Places Blue and Purple

Figure 11: "Pioneer Peak" purple and blue acrylics on canvass, 1992, courtesy of Tim Troll, Anchorage, AK. Kate climbed this Alaskan mountain as well.

A dilemma: If I voted in Alaska, my vote for John Kerry would get washed away by the conservative tide reelecting President George Bush. With the Iraq war in total meltdown, this did not sit well with me. Furthermore, if I voted in Alaska I would have to choose between two poor choices in the US Senate race. Having an apartment in Seattle (part of my work with the MSC), I realized I could change my voter registration and experience voting in a more progressive state. Why not continue to change everything?

Changing my voter registration meant I'd no longer be eligible to receive at least the next two Permanent Fund checks, an annual distribution of oil revenues to State of Alaska residents. Would it be worth at least $2,000 to experience voting *for* someone instead of the lesser of two evils? I knew the answer even as I formed the question. Besides, the 2004 Washington governor's race between Republican Dino Rossi and Democrat Christine Gregoire was shaping up to be a close election.

This race gained national attention for its legal twists and extremely close finish. Republican Dino Rossi was declared the winner in the initial count and again in the subsequent recount. Then, a legally mandated machine recount reduced that lead to only forty-two votes. Then, a hand count requested and funded by the state's Democratic Party gave Gregoire a ten-vote lead. It wasn't until after a State Supreme Court ruling allowed several hundred ballots from King County to be included that Christine Gregoire finished victorious by 129 votes.[52] When I learned that the final ballots counted included new registrants from the precinct where I lived, I knew my vote had been among those 129 votes.

Governor Gregoire went on to win re-election in 2008. In March 2008, she signed legislation making Washington the fourth state to adopt comprehensive limits on global warming, and the first state to develop a plan to train the state's work force for the transition to clean energy.[53] In 2009, *Greenopia*, the nation's largest independent directory of eco-friendly businesses, rated Governor Gregoire the fourth greenest Governor in the country.[54] Through her eight years of commitment and leadership on clean energy she became a true climate champion, culminating in the Western Climate Initiative, a regional approach to reducing greenhouse gas emissions.

Foregoing two thousand dollars to vote? Was it worth it? My one vote for her turned out to be the most significant vote in my life. When Governor Gregoire stepped down in 2012, Representative Jay Inslee stepped up to become governor. He has continued to make Washington State a leader on climate action.

In October 2013, Governor Inslee signed the Pacific Coast Action Plan on Climate and Energy.[55] The plan grew out of the Pacific Coast Collaborative, which consists of California, Oregon, Washington and British Columbia. The four governments have agreed to work together to count the costs of carbon pollution. California and British Columbia will use the carbon pricing programs they already have in place; Washington and Oregon intend to create programs. In April 2014, Washington's Governor Inslee signed a carbon reduction/clean energy executive order which directs environmental action across state agencies and mandates the development of more climate-friendly practices.[56] Yep, my $2,000 vote continues to pay off.

When one considers the economic region covered by the Pacific Coast Plan for Climate and Energy, the efforts of Governor Inslee and others rise in importance. The three states and the Canadian province have a combined population of 53 million people and a combined Gross Domestic Product of $2.8 trillion, making the Pacific Coast Collaborative the world's fifth largest economy.[57]

It is most heartening to see that states and regions were not waiting for Congress to address climate change. They were taking matters into their own hands to put a price on carbon and to engage their economies toward a path of low carbon. When signing the Pacific Coast Plan for Climate and Energy, Governor Inslee said, "We're the first generation to feel the impact of climate change and the last generation that can do something about it".

I agree wholeheartedly with the urgency tone in Governor Inslee's call to action. I'm still trying to get the new Administration (now an Independent Governor teamed with a Democratic Lt. Governor) to get engaged in the issue; to join in on the Pacific Coast regional plan; to commit to leaving all Alaska's low-quality coal in the ground. There are many compelling social and economic reasons why Alaska should engage, let alone the overwhelming moral responsibility to maintain traditional Alaskan cultures.

We also have a moral responsibility as a high per-capita polluting state to contribute to the solution. On a per capita basis, in 2010, Alaska had the third highest emissions of any state, even behind heavy oil producers Wyoming and North Dakota.[58]

By now you are probably not surprised to learn that "carbon pollution follows party lines". A recent review of statewide carbon emissions from the Yale Center for Environmental Law and Policy, found that party lines related to both carbon production and carbon consumption. "In other words, Republican states produce more carbon as a byproduct of their economies and their citizens consuming more carbon through transportation, residential and commercial activities that shape their lifestyles."[59] In other words, red states are not only bad on inaction, but also in carbon polluting. A double whammy.

A different path for millennials lies in states like Vermont, the state with the most LEED-certified (a high energy efficiency rating accepted by architects) buildings on a per capita basis. In 2011, Vermont was the number one green state as rated by *Greenopia*, the only independent rating system for eco-friendly businesses and services. *Greenopia* rated each state's eco-friendliness similar to the system used to compare businesses. Their 10,000 data point survey of all fifty states assessed greenhouse gas emissions, energy and water consumption, air and water quality, recycling rates, renewable energy generation, the number of eco-certified buildings, the number of green businesses, and how progressive their legislatures have been in adopting green measures.[60]

As noted previously, top honors went to Vermont. The other nine states rounding out the top ten included, in order of performance, New York, Washington, Oregon, Minnesota, California, Nevada, New Hampshire, Massachusetts, and Maine. Alaska finished 44th.[61] All ten states voted blue in the 2016 and 2012 presidential elections. At least, the time I got to vote blue, I was not only able to vote green but ended up casting an instrumental vote for a climate champion which then led to another.

Behind Washington on *Greenopia's* scale is California, a blue state that is greening in important ways. According to Green Tech Media, the Golden State is at the top of everyone's list as the smartest state for energy grids. "On the policy side, regulators are out in front, pushing new smart grid practices. For their part, the state's three big utilities have each developed best practice studies and frameworks that can help the rest of the country grasp the benefits of smart grid improvements."[62] In other words, in California energy technology is starting to merge with internet technology. According to award-winning journalist, Thomas Friedman, this would be revolutionary.

In his book, *Hot, Flat and Crowded*, Mr. Friedman concludes:

The Energy Internet has the potential to give us more growth with fewer power plants, better energy efficiency, and more renewable energy, like wind and solar, by smoothing out the peaks and valleys in energy demand. If we could just add another breakthrough on top of that— inventing a source of energy that would give us abundant, clean, reliable, and cheap electrons to power this Energy Internet and that would dramatically reduce our usage of coal, oil and natural gas—the revolution would be complete. Then you would be feeding clean electrons into an energy-saving smart grid, in a smart home, and into a smart car. Then when this happens, it will be the great energy transformation. It will be like two giant rivers coming together – the IT revolution and the ET revolution.[63]

If Mr. Friedman is right about this energy breakthrough then the question becomes, "Will the revolution in energy and internet technology come soon enough?" Here too I park some hope in the millennial generation. According to David Burstein, the author of *Fast Furious—How the Millennial Generation is Shaping Our World,* "Millennials see their lives as deeply interconnected with technology. The maturation of the millennials has occurred along a similar timeline and in concert with the maturation of the digital world... The actual 'dawn' of the digital era occurred between the late 1980s and the early 1990s, when most of the millennial generation was very young and some not yet born."[64] Burstein refers to the millennials as "The First Digitals".

To understand the degree of difference between my world and the world of the first digitals, Mr. Burstein quotes Steve Emerson, former president of Haverford, "As children of the Internet, they function as parallel processors, making their way by creating a personal mosaic of information and insight that they have integrated and assembled into the world as they see it. They do all this on the fly, much the way a web page gets assembled by databases and the way the Web itself provides many pieces of larger pictures. My generation, on the other hand, seems to prefer clear authorship and packaged sets of ideas that are of a piece and complete."[65]

I agree with Mr. Emerson's assessment of my generation's preference. Hence I have written this book—to package a set of ideas. Some of the ideas are mine and built around my personal story, but many are not. By now, you see I am also packaging the ideas of the United Nations, the European Union, Al Gore, Gus Speth etc. and I hope that in my packaging you find a mosaic of personal information and insight. May this mosaic flow like those millennials who assemble web pages on the fly.

Neil Young's song *Who's Going to Stand Up?* is streaming through my computer speakers right now. If I was a First Digital you would hear this song as you read these words. But since I'm not capable of such technological feats, please imagine Neil Young's hauntingly high voice with full orchestral accompaniment singing:

> *Who's gonna stand up and save the earth?*
> *Who's gonna say that she's had enough?*

Who's gonna take on the big machine?
Who's gonna stand up and save the earth?
This all starts with you and me

Neil Young's lyrics are as invocative of environmental truth today as his lyrics to "After the Gold Rush" were forty years ago: "Look at Mother Nature on the run in the ninety-seventies. Look at Mother Nature on the run in the ninety-seventies", Young whines. Delivering this message through the path of YouTube and beyond was inconceivable in the ninety-seventies. As we approach the twenty-twenties, who will take on the Big Machine? Who is going to stand up for the earth?

Answer: We all are. It starts with being political and works best in places blue and beyond.

Burstein notes that the millennial generation is the first generation to grow up with a constant awareness of climate change. Time for the last generation that can do something to team up with the first generation to grow up with climate change.

Chapter 11 — Journey into Renewal

Figure 12: On the summit of Mount Kilimanjaro, accompanied by fellow climber Laurie Normandeau and their porter. Photo taken 2004.

Shortly after turning fifty, I noticed a flyer in a Seattle coffee shop. It was seeking adventurers to climb Mount Kilimanjaro as a way to raise funds and awareness for the Fred Hutchinson Cancer Research Center. Pondering this poster and the striking images of Kilimanjaro towering above the African plains, I felt the stir of adventure start in my toes, run up my spine, and set my brain buzzing. I could do this as a way to honor the three good friends I'd lost to cancer. I could make Kilimanjaro my "turning fifty" event and an exercise motivator.

Each climber needed to raise a minimum of $10,000 before they would be accepted for the climb. It was the cancer center's way of raising funds for the researchers whose pioneering work on cancer was still too far out to attract traditional funding sources. The cancer center sought to match the cutting edge spirit of the researchers with an out-there way to raise funds. To ensure the financial commitment of prospective climbers, each climber would have monthly goals to be secured by their personal credit cards. If I didn't reach my goals, my card would be charged—creative fundraising with a serious hitch.

Twelve years later, I have discovered there are many creative, innovative ways to generate funds for a worthy cause. The notion of crowd sourcing funds didn't exist in 2002. Now crowdsourcing.org provides access to over twenty-eight hundred crowd sourcing programs for a diverse array of ideas, many of which have sustainability links. Online philanthropy marketplaces like *Donors Choose* and *Changing the Present* offer an opportunity to selectively support a wide variety of projects promoting sustainability. As a test, I recently funded one hour of a researcher's time to build on current efforts in addressing sustainability issues related to palm oil, cotton, soy, sugar, coffee and other commodities. I'm told that the researcher will determine the key impacts that matter most for water, soil, biodiversity, toxicity and greenhouse gas emissions. With this type of connectivity at my fingertips, I no longer need to bump into an alluring poster before getting on a purpose-driven campaign.

In her TED talk, "You are the Future of Philanthropy," Katherine

Fulton, president of the Monitor Institute, explains how the perception of global philanthropy is changing. "I'm hopeful because it's not only philanthropy that's reorganizing itself, it's also whole other portions of the social sector, and of business that are busy challenging 'business as usual'. And everywhere I go, I feel that there is a new moral hunger that is growing. Words like 'natural capitalism,' 'venture philanthropy' and 'philanthroentrepreneur'—we don't have a language for it yet. Whatever we call it, it's new, it's beginning, and I think it's going to be quite significant."[66]

And indeed it has been. Just looking at one global company and bank, there are over 7.2 billion euros invested in climate philanthropy.[67] Launched in September 2015 by the UN Sustainable Development Goals, the bank BNP Paribas has committed to allocating 15 billion euros by 2020 to the renewable energy sector.[68]

All I knew when I set out for Kilimanjaro was that I was some sort of "philanthroclimber" setting out on an African adventure. It was not until we met at the trailhead that I learned I would end up giving much more to the nearby villages of unemployed young men. Sitting on the surrounding fences were at least a hundred young men, each clambering to be chosen as a porter for the climb. Our lead guide opted to select fifty porters for our trip.

Fifty porters! This seemed excessive for one climbing party. I began to balk at having so many. The lead guide quietly pulled me aside to tell me about the living conditions of these men and how the food and tip money we shared would make a significant difference to each of them and their families, regardless of how many we divided it among. Humbled by the situation, I gladly passed off my load and headed for the trail.

Our route up to Uhuru Peak, the 19,344-foot summit of Kilimanjaro, followed the Machame Route, the preferred route for acclimation and success. However, most climbers take the Marangu Route, which gets them to the summit a day sooner. Also known as the Coca-Cola route for its great popularity, the summit rate of climbers on the Marangu route is half that of the Whiskey route, the Machame Route. Clearly at high altitudes it is better to age with the mountain. Kibo, the African name for Kilimanjaro, is only a thousand feet lower than Denali, and altitude sickness is a real possibility. Also on top is a small glacier in fast retreat.

Those "Snows of Kilimanjaro" Hemingway wrote about in 1936 are long gone, shrunk by a whopping 82% since the first survey of the summit in 1912. Even since 1989, when there were 3.3 square kilometers of glacier, there has been a decline of 33%.[69] At that rate, say the experts, Kilimanjaro will be completely ice-free within the next decade or two. I remember hearing the same forecast when I was climbing Kilimanjaro.

We first headed through the damp cloud forest, an enclave of vein-laden trees complete with monkeys. The cloud forest was the first of five distinct ecosystems through which we would hike. From the forest to the moorlands, the alpine desert, and up to the arctic conditions at the summit, Kilimanjaro is one of the most fascinating places in the world to hike. Every day was totally different. And every evening was filled with the same warm hospitality and service of the porters.

The two porters assigned to me out-charmed the cloud forest. They brought my bags to my tent and then offered warm water and a washcloth so I could clean up before dinner. Best of all they delivered big, wide smiles. I remember sharing music on my headset with them, lots of laughter.

The porters filled our evenings with good cheer and songs all the way to our last camp, Barafu, at an elevation of 15,200-feet. This camp was just below the steep 1,400-foot climb to the summit ridge. The two guides (one local) were briefing us on what to expect, wear, and bring. At midnight we were to leave camp and don our headlamps for the climb up. By starting at midnight, the strong climbers would be on the crater ridge as the sun broke open over the plains below. Only certain porters would accompany us and we would be divided into groups.

It was a solemn night, not just because of the briefing but because each of us was doing this climb to honor friends and loved ones who either had not made it or were struggling with cancer. I had brought them along with me.

I was in the first group of four climbers. Below me on the steep twisting trail, I saw a snake of headlamps shimmering against the dark of night. Then, as the sun peeked slowly above the horizon, the snake faded away into a splendid sunrise. Cresting the ridge and watching the glacier light up in amber glow, I felt strong on this landmark, on this majestic mountain

At the summit marker, after taking a few photos, I stepped down for the next climber and took in the moment while feeling so strong. Not even a headache. I sought out a place to be alone. I pulled out the photos of Jan, Candia, and Eric.

They came from three distinct parts of my life. Jan was my best college bud who was constantly falling in and out of love. Candia was one of the first friends I made upon landing in Ketchikan. She loved to kayak. She was the "granola ethic" personified in how she lived; yet breast cancer caught up to her in her early thirties. Eric is my friend from the Juneau part of my life. He brought creativity and good energy into all the ways that I knew him.

With moist eyes, I re-packed the photos into the side pocket of my pack. Before departing, I wanted to get close to the glacier. The low angle of light was carving an interesting pattern. Stopping to take in this curious sunlit view of the remnant ice cap, the presence of being in the moment of it came rushing in around me. The moment of summiting, honoring my friends, the beauty, it all came crashing in upon me. I turned into a lump of sobs. Something out of character for me.

After collecting myself, I could see that the path to the base of the ice cap had opened so I ventured closer. I was struck by how different it was from the robust, powerful glaciers at home. It looked shrunken and weathered and there were no melting ponds of water at the base. It looked dry, with fat fingers of ice still standing around the outside. How does a glacier shrink without melting?

The answer, I learned from the porter who came to escort me back to the ridge, is sublimation. Sublimation is the transformation of a solid into a gas without first becoming liquid. The glacier was becoming water vapor in the dry African air without ever melting into liquid water. Sublimation leapfrogs the liquid stage, and that is why the ice looked so strange.

We can learn from sublimation. We need to leapfrog across inefficient forms of energy; go from coal, bypassing natural gas, straight to renewable energy.

Renewable energy has already achieved "grid parity" with coal-generated electricity in many areas around the world. This means that the developing technology produces electricity for the same cost to ratepayers as traditional technologies. In many regions, renewable energy is now becoming cost competitive with fossil fuels. According to Al Gore, "This cost-down curve will push the price of renewable electricity to a level equal to or below the grid average price in regions of the world where 85 percent of the world's people live."[70]

Crossing over to grid parity is like ice becoming water: a difference of one degree. By comparison, one degree can make a transformative difference. In the context of renewable energy, this degree change is within reach, particularly if the energy trajectory goes from coal and oil directly to renewables. The middle step of electricity from natural gas can just be sublimated like the shrinking glacier on Kilimanjaro.

After snapping one last photo of the ice cap remnant, I dutifully hit the trail down to Barafu camp. On the way down, I encountered the last person from our group still going up. Her name was Kenneth and she was carrying a necklace of gold-plated, name embossed medallions for each person she was honoring. One member of our group had turned back and she was thinking of doing the same.

When I returned to camp, I had time to rest before packing up. Reflecting back on the moment when I had broken down sobbing, I asked myself, *what was that all about?* It had to be something deeper than summit elation or a moment of sadness missing my friends. Then it dawned on me. I've always had some inner doubt about whether or not I would have made the summit of Denali, regardless of the fierce storm that had stopped me. I remember having a headache and starting to struggle keeping food down. Here at just a thousand feet less and twenty years older I felt simply great. Surely, I would have summited Denali. That's what had overcome me. My inner doubt had lifted. I had reached two summits in one.

I fell into peaceful slumber. Two hours later, Kenneth woke me up with the news that she had reached the summit; a great day in many ways.

More than a decade later, the inspiration from "two summits in one" remains a strong memory. I keep it close to me and only use the imagery when I am so moved. I want us all to reach the twin summits of fossil-free energy and lower CO_2 emissions. To answer, "Is this possible?" I must turn to Al Gore.

Twenty-four hours before the start of the 2014 People's Climate March on New York City, Al Gore posted twenty-four reasons one can remain hopeful for a fossil free future. He did this as part of the Climate Reality Project, a non-profit educational program he set up after his documentary, *An Inconvenient Truth,* made the movie rounds. The number one reason he posted—renewable energy projects are growing in use and getting cheaper all the time. Remember the significance of wind and solar reaching grid parity with fossil fuels? As Gore notes, "It's the difference between ice and water. And it's the difference between markets that are frozen up, and liquid flows of capital into new opportunities for investment."[71]

To find out what the other twenty-three reasons for hope are, see the Appendix C. His list is deserving of the notion of reaching two summits in one.

Mr. Gore's fundamental basis for optimism comes back to the basic rules of how the world operates. In the *Rolling Stone* interview, "The Turning Point," he says, "It is worth remembering this key fact about the supply of the basic fuel: enough raw energy reaches the Earth from the Sun in one hour to equal all of the energy used by the entire world. Where most of the world's people live and most of the growth in energy is occurring, photovoltaic electricity is not so much displacing carbon-based energy as leapfrogging it altogether."[72]

Leapfrogging, like water going from a state of ice to a gaseous state without melting first. Leapfrogging, like the sublimating glacier on Mt. Kilimanjaro.

Chapter 12 — The River Knows Best

Figure 13: Kate floats by icebergs the size of apartment buildings. Alsek Lake, 2010. Photo courtesy of Bill Hanson, Juneau, AK.

I was with Bill, my husband, climbing a grassy point to scout the Channel of Death, the chaotic and always changing river entrance into Alsek Lake. The combined Alsek-Tatshenshini Rivers and Alsek Lake lie completely within the largest designated wilderness in the world. In the U.S., the Alsek River is part of both Wrangell-St. Elias National Park and Glacier Bay National Park, all told encompassing nearly 5,000 square miles of rugged mountains and coastlines. It is the ultimate wilderness and here I am with my life's companion doing something we enjoy…scouting a river. I felt home again.

Scouting allows one to assess the river's character and how to inject judgment into the pull of raging water. If not for scouting, Lewis and Clark would never have found their way around the Great Falls on the Missouri River, nor would they have made their way down the Columbia River. It is the Lewis and Clark story, our American odyssey that makes scouting such a noble endeavor.

The 147-mile Alsek River pours from its headwaters in the Kluane mountains of the Yukon Territory southward through Turnback Canyon, a dangerous and seldom-run stretch of class five rapids, and then across the imaginary line that marks the border between Canada and the United States where it is joined by the braided gravelly channels of its tributary, the Tatshenshini River, increasing its volume threefold. Farther downstream the mighty Alsek, with flows that are well above 80,000 cubic feet per second, takes a hard left turn around a short mountain to flow easterly into Alsek Lake, and then out of the lake again to make its run south to its mouth in Dry Bay on the edge of the Gulf of Alaska.

Alsek Lake is fed by two gigantic glaciers. The Grand Plateau Glacier descends Mt. Fairweather—at 15,300 feet it is one of the world's tallest coastal mountains. The second is the massive Alsek Glacier. The lake is notorious for house-sized icebergs and strong winds.

The extraordinary setting of this wilderness beckons the memory of Lewis and Clark. Except that back in their day, they were making the map as they went. Here on the Tatshenshini-Alsek river system, I had a guide book and experienced river runners to consult with. I even had a waterproof map naming all the danger points. But with changes in water level, rearranging gravel bars, and transformations in log jams and stray trees, the river is different each and every day.

The Channel of Death is not named for lives lost, but for the possible consequences of casual approaches. So says the river guide to this ice-laden entrance. With purpose in our step, we climbed a low hill, perhaps a last remnant of bedrock from millennia of scouring by glacial ice, perhaps a moraine of gravel and rock deposited by the retreating glacier. Below us, our friends fanned out across the sandy flats and shoreline. Beyond them floated an expanse of wind-driven ice that stretched for more than a mile into the interior of the lake and more than a mile toward the far shoreline where the river flowed out again. We would reconnoiter, compare our reports, and then decide whether to make a run for it immediately, or if prudence would tell us to set up camp and wait for the shifting ice to open a channel through which we could safely row our rafts.

My twenty-five-year-old son, Rion Alsek Hanson, was along for our ten-day journey down the Tatshenshini-Alsek Rivers. Yes, this is his namesake river and would be the first time down it for the three of us. It was, however, not our first wild river trip together. When Rion (pronounced Ryan) was eleven years old and his sister, Erin, was thirteen, our family spent a week kayaking down the Stikine River, another glacial river that breaks through the Coast Range, 165 miles from Telegraph Creek, British Columbia, to the mouth near Wrangell, Alaska.

Through decades of running rivers I have come to appreciate the teaching moments that a river can offer. Here is a snapshot of a few of those lessons.

Reading and Running a River: The fastest way down the river is not always the safest. Reading a river teaches us how to interpret key information and to be ever aware of the present …but always with a long view as to what's happening down river, i.e. reading the present with anticipation of what's around the bend.

I have been dumped into cold, glacier-fed rivers two times, once from a canoe and once from an inflatable kayak. The time I flipped my inflatable kayak I was casually trying to avoid the rapids straight ahead. I was already wet from the previous set of rapids and didn't want to take on more water. I opted not to figure my way through the rapids and paid for it with a sudden dunking.

Negotiating Danger Points: Once you know your route through the rapids, you must commit to it. Once you're in the rapids, you read the flow. Negotiating rapids teaches us to find the balance between being decisive and going with the flow.

While rapids are fun and adventuresome, the real value of river trips comes when there are days upon days of just going downstream through the mountains, the forests and the flats. The changes in the passing landscape allow one to appreciate and wonder about one's place in the greater world.

When one takes the time to **Understand the Natural History of a River,** one learns about man's ecological relationship with the river. Furthermore, reading about natural history teaches one about the balance of a watershed and how best to live within it. Then as days go on, floating by cliffs, by eddies of calm waters, the natural history knowledge ferments into a sense of place, of belonging.

Many of my river trips came in between jobs as it was a way to take stock and think about my professional path. It was on these in-between-jobs river trips that I realized being an effective leader means gaining the strength of others like tributaries feeding into the river, keeping the flow strong and steady. Instead of trying to blast through obstacles, it is better to negotiate around them with the help of rising waters.

Whenever I was **Contemplating a River**, I learned something new about life and my place in it. I learned that slowing down and watching the never-ending flow teaches that the beauty is in the journey. The contemplative river also teaches the value of "interconnectedness" to all living things. Furthermore, because the river is never stationary, always changing, one gains a sense of impermanence and learns the value of letting go of expectations, of being attached to set outcomes

I am still working on this last lesson. Actually, I hope to be working on it until I die. In the meantime, a narrow channel appeared to be open as far as we could see into the mass of floating ice. I wanted to paddle through it so we could camp for two days along the far shore of the lake where the Alsek River flows out. Fred, our head oarsman who had run the river before, counseled against this approach. He warned that there could be a lot of ice out of view. If we attempted to go straight through the pack, it might take extra hours or even prevent us from crossing. This wasn't a matter of simply rowing across a lake. The enormous Alsek River flows into the lake and then out the other side. If the ice blocks the channel, the entire river dives below the ice jam. The situation changes by the day and many rafters have had to abandon their trip here with a pick up by float plane. Knowing that there would be no turning back once committing to the channel, we all agreed to set up camp and rest for an early morning departure.

The following morning, the wisdom of Fred's counsel became apparent as soon as we entered the narrow channel between the ice and the shoreline and heard the rush and roar of water from the giant river submerging beneath the ice pack opposite us. Broken clouds revealed large islands of ice. In the distance we could see the icefalls of Alsek Glacier. Far out into the lake we saw what appeared to be an opening around the mass of ice islands. Our only option was to ride the current along the shore, and then row far out into the lake where we might be able to go the long way around the ice pack. Floating past spears of ice, absorbing the magnitude of the blue towers, the vast glassy lake, the glaciers, my spirit soared upward.

John Muir taught us that the path to the Spirit is not away from the world, but deeper into communion with the primary forces of nature where Spirit is lightly clothed. Here, in the world's largest area of protected wilderness, I was surrounded by the primary force of glaciers and the spirit of beauty reflected in calm, cold water. It is in these moments that I feel most interconnected to the world at large.

A distinct pyramid of ice floated ahead. We approached and found an equally perfect pyramid in reflection. I smiled inside and was reminded of what the Buddha said about perfection: "When you realize how perfect everything is you will tilt your head back and laugh at the sky."[73]

As our raft approached a new lead of open water around the next set of grounded icebergs, I saw a deep blue hole in the center of the ice pyramid. I was transfixed looking into the womb of ice that carved mountains. The closer I got to the hole, the more cracking, groaning, and moaning I heard. The whole of nature was alive through ice.

Taking a turn at the oars, I eventually saw open water all the way to a gravel beach on the far shore at least a mile away—our long sought campsite for our last night on the river. We were alone in a vast landscape. The clouds began to lift. A small opening of blue appeared above the campsite in the distance. The sight of it powered us toward shore.

We set up camp quickly and ate a chicken fajita dinner complete with pico de gallo. This is food Lewis and Clark could not have fathomed eating while in the wilderness. They were far more intrepid, relying on food they could hunt and gather. Nowadays we even had folding chairs to rest on. With dinner done, we lined our chairs along the lake's edge to take in the evening's ice show. A soundtrack of colliding, reverberating ice was already underway.

Sipping the last of our wine we sat for hours watching the buildings of ice reflect the evening light split by clouds. Moving ever so slowly, fountains of ice drifted toward each other. Every so often we heard a sharp rumble in the distance. In unison we would turn our heads toward the faraway glaciers, but we never saw the ice fall. The glacier was miles across the lake. By the time the booming crash reached our ears, we could only catch a glimpse of the remnant waves and splashes.

Closer by, we found unexpected pleasure in the chain reactions caused by the ice thunder echoing across the lake. The expanding ripple of long-ago calving eventually reached us, rocking icebergs back and forth, sometimes causing them to flip over. More ripples. Then calmness would reign again.

We arose from our chairs to applaud this spectacular ice show made possible by climate change and two rapidly retreating glaciers. The irony was not lost on us, but the moment won out, deserving group praise.

These moments brought perfection to our last night of camping together. In the morning we had only a three hour ride to our take-out point, Dry Bay in Glacier Bay National Park. Drifting down the gray river, the glacial silt hissing against the hulls of our rafts, swirling fog and floating icebergs and one last set of rapids, we bade farewell to another therapy session with the river.

Several years after our journey down this wilderness river, I learned that the Alsek River is one of the few rivers in the world classified as an "antecedent river." This is a geological term for a river that predates the rising of the mountains it currently cuts through. For this to occur, the rate of erosion must match the rate of mountain building. This allows the river to hold its course, a rare balancing of forces between rivers and mountains. It is this balance that humanity must find. We must hold the rate of population erosion constant while building a new economy that factors in the needs of humanity with the ecology of life. This is a mountain of a challenge. But we are a country with a history of overcoming big challenges. We are a country with a history of setting about on grand adventures.

In the PBS Historical Series, "Lewis and Clark: The Journey of the Corps of Discovery" by Ken Burns, historian Stephen Ambrose tells us that "the number one human lesson of the Lewis and Clark expedition is what can be accomplished by a team of disciplined men who are dedicated to a common purpose. Teamwork. The number one story here is there is nothing that men can't do if they get themselves together and act as a team."[74] Without detracting from this powerful statement, I must point out that the series also acknowledges the critical role that Sacagawea played in their good fortune and eventual success.

The map produced by Lewis and Clark was the product of a courageous expedition that forever expanded the frontiers of our continent and our imagination, of what can be accomplished by joining together in a grand purpose. Today, the grand purpose is finding a way to bring human civilization into a sustainable relationship with the surrounding environment. We no longer need to subdue nature and find routes through the wilderness so that human civilization can prosper. Now it is the other way around. Human civilizations need to find new ways of living with Nature so both can prosper.

How do we do this? To answer this question, read Gus Speth's *America the Possible*. Also there are other great environmental thinkers like energy guru Amory Lovins, economist Paul Hawkens, think tank founder Lester Brown, and journalist Tom Friedman, all of whom have charted policy direction for a sustainable future. We do not lack vision. The challenge lies in coming together to channel all our energy and knowledge in one direction, much like the hundreds of streams that feed into the Alsek River.

My time on the Alsek River taught me one thing more. From circumnavigating Alsek Lake to the point where the river comes together again, I learned that Being One with a River allowed me to bring my personal awareness into the presence of the river while learning my place in it. Being in tune with the river taught me when to inject judgment or mindfulness into the pull of life.

Figure 14: Certificate of Appreciation for Climate Change work, signed by former Governor Sarah Palin when she did believe in the science of climate change.

It wasn't until I took the position of executive director of the Alaska Conservation Alliance and Alaska Conservation Voters that I knew I was done seeking higher policy positions with the State of Alaska—that I was okay with just going with the pull of life. Though the main current eluded me, I was still on a side-channel to policy-making. Only this time I would be wearing a straight-up green hat.

It's one thing to be a greenie in a fish hat, it's another matter completely to be out in front. I had to make a concerted effort to convince legislators and other state leaders that although my hat had changed, I was still the same pragmatic Kate who cared about the economy as well.

About the time this new hat was wearing comfortably, the pull of life crashed into my professional world. The crash came in the form of a 6 a.m. phone call on August 29th, 2008 from the National League of Conservation Voters screaming, "It's Governor Paul-in!"

"Oh, you mean Governor PAY-lin," I retorted to the League's media director, who wanted my input on a response to Senator John McCain's naming of Sarah Palin as his running mate in the 2008 elections. Specifically, the media director wanted to know about Sarah Palin's conservation record. I started to describe our educational strategy on renewable energy and how Governor Palin supported our efforts to establish a renewable energy grant fund when she stopped me short. "We'll take it from here." she informed me.

Once the magnitude of Senator McCain's gamble on Governor Palin sank in, I realized that her candidacy was far bigger than my working relationship with her. I needed to gear up for the national debate on climate to ensure that my organization had an ability to participate in the unfolding story of introducing Sarah Palin to a national audience. The next day I called two staffers to help me put together a media packet of video clips and articles on Sarah Palin and conservation issues. We were fortunate in that Sarah Palin, as a candidate for governor, had attended the first-ever gubernatorial debate hosted by Alaska Conservation Voters. We had video of her talking, a sought after commodity. On the Monday following the big announcement, the newspaper calls started coming in.

In one day, I handled at least twenty newspaper and blog inquiries about Ms. Palin's environmental record. I offered them a media packet that included a conservation issues paper. It was well received and passed around. More interview requests came in. In short, I acted on a strategy that's in every coach's playbook, from youth soccer to professional football: **seize the moment.** In the course of giving interviews I even came up with a witty line for being picked up. When asked what I thought about Governor Palin's environmental record, I responded, "When it comes to the environment the only difference between George Bush and Sarah Palin is lipstick." Yes, I seized the moment.

Here's another insight from those days of political roller coaster riding. It comes into light through a media inquiry from the L.A. Times. Sitting in a loveseat overlooking Gastineau channel, watching fishing boats out my window head out to fish, the phone rang. It was a reporter, Ms. Kim Murphy.

"We're doing a story about Governor Palin's first six months in office," she said. *"We petitioned her office for her appointment schedule for that time period."*

"What does this have to do with me?"

"You were the only non-oil and gas person on her schedule. What did you talk about?"

I paused, and then said, "Basketball."

"Basketball?! As the executive director of the Alaska Conservation Alliance, you talked basketball?"

"You know the importance of making a connection?" I queried. "She was a point guard in high school and so was I. Once she felt comfortable with me, I was going to make my ask."

"Which was what?"

"That she and certain members of her cabinet meet with select

members of the Alaska conservation community."

"And did she give you that meeting?"

"I got a full hour meeting with her and two of her most influential commissioners, all fifteen of us," I said.

I could hear exasperation on the other end of the phone. "Getting a meeting doesn't really make for a story, does it?" I suggested.

"Not really." Click. Ms. Murphy had another call to attend to.

While there wasn't a good newspaper story here, it was important strategically. Our introductory meeting with the new commissioners of natural resources and environmental conservation went quite well. A dialogue had been established. In essence, **connect first, then make your ask.** To ensure a solid connection with Governor Palin, we left behind climate change petitions signed by evangelical ministers and talked first about oil taxes, her signature campaign issue. Governor Palin rode into office on a white horse of tax reform and ethics.

Prior to her half-term as governor, the Alaskan political scene had been mired in oil tax corruption and mud. A number of legislators bribed by Veco Corporation, an oil support company for Alaska's North Slope producers (BP, Exxon, and Conoco-Phillips), even had a name for their ring of legislators and lobbyists: *The Corrupt Bastards Club.*[75] They were caught on FBI surveillance video sporting baseball hats with CBC across the front. The scandal ran deep and radical change was needed. Former Governor Walter Hickel, known for his "owner-state" philosophy when it came to oil taxes, stepped into the spotlight. He didn't want to run again. He threw his political and financial weight behind the plain-speaking mayor from Wasilla: Sarah Palin. She handily beat the incumbent Republican governor Frank Murkowski in the primary and zoomed easily into the governorship. Knowing this background, it would have been disrespectful as an Alaskan not to talk about oil taxes before asking to set up a climate change sub-cabinet.

It's significant to note that prior to my meeting with newly-elected

Governor Palin, I had been making the rounds to local Rotary Clubs and Chambers of Commerce with a slideshow and talk about climate change in Alaska. Not only had I warmed up her audience, but these talks greased the opportunity to meet with the new governor within the first month of office. She even appointed me to her transition team for the Department of Natural Resources.

If your aim is to influence decision-makers, you need to go where they go, and you need to go in person. This is all part of the strategy to conduct public outreach outside the conservation community.

Social media and email are wonderful for generating attention during key votes or elections, but they don't establish the type of relationships decision-makers feel most comfortable with. Go to where your adversaries meet, whether it be a business luncheon, spa, or a ball game. If possible, let them get to know your reasonable side first. For example, before doing my climate speaker's tour of local Chambers of Commerce, I did a round of Chamber talks featuring my eco-nomic principles. Since that talk went over well enough it was a much easier ask to get a speaking opportunity on climate change, a topic they were not warm (pun intended) about engaging in.

When approaching audiences of a non-green persuasion, I've learned it is important to start off on friendly, familiar territory. For example, my PowerPoint talk on climate change started with these quotes from Big Oil[76]:

> *Human activity, including the burning of fossil fuels, is contributing to increased concentrations of greenhouse gases in the atmosphere that can lead to adverse changes in global climate.*
> —Conoco Phillips, 2006

> *When 98% of the scientists agree, who is Shell to say, "Let's debate global warming?"*
> —Shell, November 2006

> *Recognizing the risk of climate change, we are taking actions to*

improve efficiency and reduce greenhouse gas emissions in our operations.
—Exxon Mobile, December 2006

Even though Al Gore's arguments and lines might be more persuasive to conservation minded people, they will fall on deaf ears in red states like Alaska. When given the chance it is important to use conservative messengers.

To set the stage comfortably it is important to use messengers that match the audience. Meet the audience where they are before trying to raise their awareness. Audiences prefer to be brought along rather that hit over the head with dire predictions from "those liberal big thinkers."

Fortunately, there are plenty of places to find conservative voices. When you approach them you need to be strategic. When the conservation community met with Governor Palin, we used a climate change petition signed by evangelical ministers to catch her attention and to make her feel at home with our subject matter. We may have had only ten ministers, but it was enough to make her sit back and listen. In the end, she agreed to appoint a Climate Change Sub-Cabinet.

When it comes to pushing for action, the best conservative messenger I've found is former Secretary of Treasury Henry Paulson. His New York Times op-ed (June 2014), "The Coming Climate Crash," starts out:

> *There is a time for weighing evidence and a time for acting. And if there's one thing I've learned throughout my work in finance, government and conservation, it is to act before problems become too big to manage. For too many years, we failed to rein in the excesses building up in the nation's financial markets. When the credit bubble burst in 2008, the damage was devastating. Millions suffered. Many still do.*
>
> *We're making the same mistake today with climate change. We're staring down a climate bubble that poses enormous risks to both our environment and economy. The warning signs are clear and growing more urgent as the risks go unchecked. This is a crisis we can't afford to ignore.*[77]

Secretary Paulson then lays out the case for a carbon tax.

> *We need to craft national policy that uses market forces to provide incentives for the technological advances required to address climate change. We can do this by placing a tax on carbon emissions.*[78]

Most conservative messengers will not be as direct and articulate as Secretary Paulson, and, like many of us, could use assistance writing op-eds. When this happens, activists who write well should seek opportunities to ghostwrite op-eds or letters to the editor for well-placed conservatives who may be sympathetic to the economic and security impacts of climate change.

Another strategy I often use in messaging is to match the environmental topic of concern with the most trustworthy spokesperson. For example, a school nurse is far more effective a spokesperson on toxic substances in our drinking waters than a conservation leader.

The voice of conservation doesn't need be front and center, just undeniably present. Wear other hats, wear the hat of the best occupation or association that fits the topic. When there is no appropriate conservative messenger to use or assist, wear the hat of a parent or a civic volunteer. There are times when wearing the green hat works and times when it doesn't. When in doubt, wear another hat—one that builds a bond with your audience.

However, there are times when the green hat works best. I was wearing the hat of Alaska Conservation Voters when we sought to defeat Governor Murkowski's natural gas pipeline proposal that would have committed $80 billion of state funds to Big Oil (Exxon, Conoco-Phillips, and BP). All the major unions were opposed to the proposal because it would divert funds away from the capital budget. State economists were opposed because the give-away terms had the State of Alaska carrying most of the risk. Then to make matters worse for Governor Murkowski's gas pipeline proposal, the very people negotiating the deal walked out. They became the nexus for reviving an activist organization called Backbone. This ad-hoc group, comprised of many unions and several small business organizations, reached all across Alaska.

After two meetings it was clear to the movers and shakers in the room that the greenies could contribute and work with a diverse group of concerned Alaskans; that we had a network of activists to call upon. Once the union representatives and business leaders saw the value of including the conservation community I was gaining allies for possibly working together on other issues. In other words, play big on non-green issues when given the chance.

If green is going to succeed at all in red states, green needs to be bigger than green. If the issue is budgetary in nature, you can usually tie it back into repercussions for natural resource management and environmental protection. If the issue is tied to clean water, you can link it to children. Or, if you are in Alaska, you can often tie the issue to salmon and commercial fishing. The options are out there. Find them and build bridges for future green related issues.

I have no doubt that the relationships I established on the ground level with labor unions paid off when seeking support for a robust renewable energy fund. Under the guidance of a visionary director, Chris Rose and the directors (including me) of the Renewable Energy Alaska Project (REAP), the State of Alaska established a $500 million dollar renewable energy fund. In the campaign to establish this renewable energy fund, and in the supporting literature, climate change was never mentioned. The greater issue just sat in the back as the silent all-knowing beneficiary of any action to bring Alaska into the clean energy economy.

If you can achieve goals related to renewable energy or energy efficiency—such as upgrading building codes—without raising hot button phrases like "global warming" then do so. When the end conservation result can be achieved without cloaking it in green terminology, then do so. Avoid hot buttons and spin your message for the greatest political effectiveness. For example, "Climate change" is preferable to "global warming" because some geographic locations will experience more snow and cold rather than warming. Push more palatable buttons first. "National security," "reducing our addiction to oil," "being part of the clean energy economy," and "reducing the cost of inaction" are all good phrases to push early in the campaign. Then educate along the way. The reality of climate change will be there in the end.

Once Governor Palin got comfortable with doing good things for renewable energy, the conservation community along with native organizations and a few new allies pushed to have her set up a climate action strategy similar to other states. Although she had already signed an Executive Order in September of 2007, to establish a Climate Change Sub-Cabinet, her administration was not really engaged in the actual preparation and implementation of an Alaska Climate Change Strategy. Now a year and half later, it was time to set up a process. Governor Palin concurred and through her Department of Environmental Conservation she set up two working groups of Alaskans to help craft this strategy; one focused on adaptation and the other on mitigation. She even appointed me along with oil and gas representatives to serve on her Mitigation Advisory Group.

Upon completion of developing a set a recommendations to reduce Alaska's greenhouse gas emissions Governor Palin issued each of us a certificate of appreciation for our "outstanding contribution to the development of a Climate Change Strategy for Alaska". She signed this certificate June 2009. She resigned as Governor one month later.

In that intervening time from when I first got that phone call announcing Governor Palin as Senator John McCain's choice for Vice-President and now, Governor Palin has done a complete one-eighty on the subject of climate change. Here is what Sarah Palin said in October of 2014 on her channel:

> *I'm not a denier. I don't doubt that climate change exists. No one has proven that these changes are caused by anything done by human beings via greenhouse gases. There's no convincing scientific evidence for man-made climate change. The climate has always been changing. Climate change is to this century what eugenics was to the last century. Its hysteria and a lot of its junk science.*[79]

While this about-face is downright shameful, my mother taught me to give credit where credit was due. To this day, I credit former Governor Palin with setting the goal of sourcing fifty percent of Alaska's electrical generation from renewable energy by 2025. Goal setting with no path forward is a little step but by giving Governor Palin big credit we were able to help set up the Alaskan legislature for passing the Alaska Sustainable Energy Act (SB 220) and the State Energy Policy (HB 306) in 2010. These two bills together put substance behind Ms. Palin's goal setting. Thus demonstrating that the strategy of giving big credit for little steps can pay off.

Remember, it's not important that you get recognition; it's important that the *issue* gets recognition even if it comes from someone you may not be politically aligned with.

Fortunately, half-term Governor Palin's view about climate change is in the rapidly declining minority. Did you know that if a political candidate makes the following statement, you will be more likely to vote for that candidate?

> *I believe that global warming has been happening for the past 100 years, mainly because we have been burning fossil fuels and putting out greenhouse gasses. Now is the time for us to be using new forms of energy that are made in America and will be renewable forever. We can manufacture better cars that use less gasoline and build better appliances that use less electricity. We need to transform the outdated ways of generating energy into new ones that create jobs and entire industries, and stop the damage we've been doing to the environment.[80]*

According to a Pew Research survey, this is the most politically acceptable statement a politician can make about climate change. A far cry away from Sarah Palin's denial rant.

Looking back on my time in and around the former half-term Governor I'm most proud that our conservation issues paper (the one my staff and I prepared upon the announcement of her Vice-Presidential candidacy) made it into Senator Joe Biden's briefing material for the Vice-Presidential

Debate. I also learned from the National League of Conservation Voters that our paper "Conservation and Governor Palin" helped them to get a climate change question inserted into the Vice-Presidential debate.

Continuing to give credit where it's due, none of this would have happened without Sarah Palin coming into my professional life. What I once thought was an interesting career side channel suddenly put me back into the mainstream.

I would like to also thank Sarah for giving me a story complete with political strategies to pass on. There is however, one more political strategy that needs to be highlighted; one I've already touched upon: always address the economic linkage. For example, in the latest climate assessment by the International Panel on Climate Change (IPCC), the IPCC refutes the claim about economic trade-offs right up front. The IPCC explains that ambitious mitigation programs and policies would not have any significant effect on economic growth and that the global economy would still be growing by 1.6 to 3 percent per year with mitigation strategies in place.[81] Putting this connection up front in a science paper, now that's a strategy I happily credit the IPCC for teaching me.

Chapter 14 — Elves Fight Back

Figure 15: Kate, wearing a wig convenes a meeting with the Board of Trolls at her sister-in-law's home; aka the Home for Wayward Baby Dolls, Morehead, Kentucky, 2013.

I am a Troll in the Tongass, the Tongass National Forest. Besides being the nation's largest forest, the Tongass is an archipelago and adjacent mountainous mainland of narrow fjords and islands, clearly resembling the folklore haunts of Norway's Trolls. All it takes is for one giant Sitka Spruce on a steep slope to fall over in a wind storm, releasing the tree's intense hold on rocks and boulders, to create a large, dark hole; a perfect hideaway. Walk deep into the moss-draped woods where the water disappears from view, listen to the haunting call of the raven, peer into the dark, wet cave at the head of a prominent valley, and tell me you don't think of the possibility of hidden folk being around.

Here hidden folk would include the Kooshdakhaa, a mythical shapeshifting creature found in the stories of the Tlingit people of Southeast Alaska. Loosely translated, the term means "land otter man." In Iceland, the term for hidden folk is Huldufólk and applies to elves. To my mind, trolls have always been cousins to elves. Now that I live in the Tongass, I imagine that trolls are related to people that resemble land otters. In any event, I was most pleased to read that in Iceland elves actually have political representation on environmental issues.

Most people in Norway, Denmark, and Sweden have not taken elves and trolls seriously since the 19th century, but elves are no joke to Iceland's residents. According to a 2007 survey, fifty-four percent of Icelanders think it is possible that elves exist.[82] In December 2013, elf advocates joined forces with environmentalists to urge the Icelandic Road and Coastal Commission to abandon a road project that ran through the middle of elf habitat. According to Associated Press reporter, Jenna Gottlieb, "The project has been halted until the Supreme Court of Iceland rules on a case brought by a group known as Friends of Lava, who cite both environmental and cultural impact—including the impact on elves—of the road project. Although many of the Friends of Lava are motivated by environmental concerns, they see the elf issues as part of a wider concern for the history and culture of a very unique landscape."[83]

Terry Gunnell, a folklore professor at the University of Iceland, said he was not surprised by the wide acceptance of the possibility of elves. "This is a land where your house can be destroyed by something you can't see (earthquakes), where the wind can knock you off your feet, where the northern lights make the sky the biggest television screen in the world, and where hot springs and glaciers 'talk'."[84] This sounds exactly like where I live.

Professor Gunnell elaborates, "Everyone is aware that the land is alive, and one can say that the stories of hidden people and the need to work carefully with them reflect an understanding that the land demands respect." Even Iceland's most famous celebrity, Bjork, defends the elves. When asked by comedian Stephen Colbert if people in her county believed in elves, she replied, "We do. It's sort of a relationship with nature, life with the rocks. The elves all live in the rocks, so you have to. It's all about respect, you know."[85]

When I think about elves as messengers for the land, demanding respect, it brings me right back to Aldo Leopold's land ethic. I am also a Troll who holds onto Leopold.

Published in 1949 as the finale to *A Sand County Almanac*, Leopold's essay "The Land Ethic" defined a new relationship between people and nature and set the stage for the modern conservation movement. Leopold understood that ethics direct individuals to cooperate with each other for the mutual benefit of all. One of his philosophical achievements was the idea that the "community" should be enlarged to include non-human elements such as soils, waters, plants, and animals.

"That land is a community is the basic concept of ecology, but that land is to be loved and respected is an extension of ethics," writes Leopold. "The land ethic simply enlarges the boundaries of the community to include soils, waters, plants, and animals, or collectively: the land. Examine each question in terms of what is ethically and esthetically right, as well as what is economically expedient. A thing is right when it tends to preserve the integrity, stability, and beauty of the biotic community. It is wrong when it tends otherwise."[86]

More than sixty years after Aldo Leopold first gave us "The Land Ethic," we find that not only do elves, as representatives of the respect that the land deserves, have legal environmental standing, but "love of place" is winning more and more battles against environmental degradation. Naomi Klein in her book, *This Changes Everything*, has a name for this rise of environmental battles and victories brought about by a love of place. She calls it "Blockadia."

> *Blockadia is not a specific location on a map, but rather a roving transnational conflict zone that is cropping up with increasing frequency and intensity wherever extractive projects are attempting to dig and drill for open-pit mines or frack for gas or build tar sands pipelines. Resistance to high-risk extreme extraction is building a global, grassroots, and broad-based network the likes of which the environmental movement has rarely seen. And perhaps this phenomenon shouldn't even be referred to as an environmental movement at all, since it is primarily driven by a desire for a deeper form of democracy, one that provides communities with real control over those resources that are most critical to collective survival— the health of the water, air and soil. In the process, these place-based stands are stopping real climate crimes in progress.*[87]

In some instances this love of place has far-reaching consequences for the climate challenge. One of the best examples of love of place spilling out into a climate victory is in the Pacific Northwest where a powerful combination of resurgent Indigenous Nations, farmers, and fishermen teamed up with community members to halt the construction of a major coal terminal in Bellingham, Washington. Listen to these words of local activist, K.C. Golden: "The great Pacific Northwest is not a global coal depot, a pusher for fossil fuels addiction, a logistics hub for climate devastation. We're the last place on Earth that should settle for the tired old retread of the false choice between jobs and the environment. Coal export is fundamentally inconsistent with our vision and values. It's a *moral* disaster and an affront to our identity as a community."[88]

You can hear something similar when looking at the lineup of opposition to the Keystone XL projects from the First Nations tribes in Canada to Texas landowners. "I just don't believe that a Canadian organization (TransCanada) that appears to be building a pipeline for their financial gain has more right to my land than I do," says Julia Crawford who is challenging TransCanada in court over their attempt to use her Texan ranch, land that her grandfather purchased in 1948.[89]

In the battles over both the Bellingham coal terminal and the Keystone pipeline, opponents are responding to the moral context of the issues. This is closely aligned with a love of place. According to communications consultant, Betsy Taylor, who in 2014 updated her messaging guide on climate and clean energy, *Climate Solutions for a Stronger America*, the moral argument is the winning argument.[90] Ms. Taylor advises climate activists to speak first about the trend of severe weather and about our obligation to protect our children from the consequences, as this touches on the underlying value of moral responsibility. Secondly, she tells us to speak about the urgency of the climate change issue, explain how we're experiencing climate impacts now, and lastly stress that the costs of inaction (the costs of waiting until we must face a disaster) far outweigh investments in solutions.

In the book, *Moral Ground, Ethical Action for a Planet in Peril,* Editors Kathleen Dean Moore and Michael Nelson posed the following question to some of the great minds and leaders of today, including Barack Obama, Carl Safina, and the Dalai Lama: "Do we have a moral obligation to take action to protect the future of a planet in peril?" While all answered with a resounding "yes," I was surprised to see how widely the spectrum of moral obligation extended.

Do we have a moral obligation to act on climate change?

Yes, for the survival of humankind.
Yes, for the sake of the children.
Yes, for the sake of the Earth itself.
Yes, for the sake of all forms of life on the planet.
Yes, to honor our duties of gratitude and reciprocity.

Yes, for the full expression of human virtue.
Yes, because all flourishing is mutual.
Yes, for the stewardship of God's creation.
Yes, because compassion requires it.
Yes, because justice demands it.
Yes, because the world is beautiful.
Yes, because we love the world.
Yes, to honor and celebrate the Earth and earth systems.
Yes, because our moral integrity requires us to do what is right.[91]

As noted by Naomi Klein, it is this expanding moral clarity that creates a new challenge for the extractive industries. It's hard to counter our responsibilities for the sake of our collective survival, for the sake of the children or, if you're in Iceland, for the sake of the elves.

Chapter 15 — Extreme Oil = Extreme Times

Figure 16: "Planet Ocean". Pastel on board courtesy of Ray Troll, © 1992, Ketchikan, AK.

When Lt. Governor Sean Parnell became governor after Sarah Palin's half-term departure, big oil finally had one of their own in office. Governor Parnell once worked as the government relations director for the oil company Conoco-Phillips, and represented Exxon in the settlement of the Exxon Valdez oil spill. Once sworn in, Governor Parnell and his administration set about to rewrite oil tax legislation and remove resource protection measures they didn't like. In essence, the Parnell Administration wore their oil logo laden jackets as badges for rollback action within the Alaska Legislature.

First up for slicing was the Alaska Coastal Management Program (ACMP). It was through the ACMP that wildlife habitat had any regulatory standing and as such, the Department of Fish and Game have been in the cross-hairs of the Alaska Oil and Gas Association for decades. Now, with Governor Parnell in office, the chicken coop was wide open for the taking and in no time wasted the Alaska Oil and Gas Association jumped in. Before long, the bill to reauthorize the ACMP failed.

As you now know this was a huge loss for me. A huge part of my professional contribution to balanced resource management had just been wiped away.

After this defeat, many searing editorials and letters followed. I joined other local community leaders to collect more than 30,000 signatures in two months' time to support the re-instatement of the ACMP on the ballot. Long story short. Although, we got it on the ballot, the Lt. Governor put it on the primary ballot and not the general election ballot. The timing could not have been worse.

Summer is a time when Alaskans pay very little attention to political issues. This gave the advantage to whichever side could dominate the air waves in August. The oil companies spent $1.5 million, compared to $200,000 raised by the "Your Coast, Your Vote" campaign.[92] Although there was no concrete evidence, the companies claimed the ACMP had been hurting the economy for the last thirty years. A flat out lie. It didn't matter they got the airwaves early and outspent us eight to one. We never had a chance. The ballot initiative to restore the ACMP went down mightily, sixty-two percent to thirty-four percent.

Next up for slicing—oil taxes. Governor Parnell introduced legislation to give BP and Exxon, two of the most profitable companies in the world, a generous tax break. The plan was sold as a way to stimulate more oil drilling and production on Alaska's North Slope and put more oil back into the oil pipeline that runs from Prudhoe Bay to Valdez. An important aside: two key state senators also worked for the oil company Conoco-Phillips. They declared their conflict of interest but were allowed to vote on the Governor's bill giving their direct employer a lush tax break that everyday Alaskans could ill afford. In 2015, when the Governor's oil tax giveaway passed, Alaska had a $3.5 billion budget gap. Yet, the State of Alaska gave big oil over $600 million in production tax credits.[93]

With Governor Parnell at the helm, Alaska soon became a vivid example of corporatocracy in action.

Corporatocracy is when both the prevailing economic and political systems used to govern a country or state are controlled by corporations and their associated interests—when corporations move beyond being the principal economic actor to also being the principal political actor as well.[94] In essence, corporatocracy is when there exists a pattern of government systems controlled by corporate interests.

As a regular columnist for the Juneau Empire, Alaska's capital city paper, and a contributing columnist for the statewide Alaska Dispatch News, I wrote about corporatocracy in Alaska. To determine if there was a pattern of undue corporate influence in Alaska's democracy, I kept track of unfolding political events within elections and the Alaska Legislature and invited the readers to count along with me. If you read the article, "Corporate Interests 4, Alaska Zero. Game over?" which is in the Appendix D you will learn that by July, 2013, corporate interests were clearly the principal political actor in Alaska's civic affairs.

If I were to add in events since 2013, the score would go up. As with the destruction of the Alaska Coastal Management Program, the citizens of Alaska pushed back against the oil tax revision with a ballot initiative. This time big oil outspent the coalition of unions, teachers, and citizen groups by a ratio of a hundred to one.[95] The ballot initiative failed, although by much slimmer margins. Meanwhile, Alaska's fiscal crisis grows as does the amount of tax credits owed to international oil companies. The corporatocracy score is at about 7 to 1 depending on how many special sessions of the Alaska Legislature one counts as single political events. Nonetheless, in Alaska we have a clear pattern of government systems controlled by corporate interests—big oil in particular.

As a side note, the score of 1 for Alaska Interests is due to a new Governor being elected. We had an Independent candidate for Governor team up with a Democrat to form a Unity Party in order to oust Governor Parnell. The newly elected Governor, Bill Walker, made a modest effort to reform Alaska's oil tax structure. Though the score of 1 for Alaska's interest is most welcome, it does not alter the fact that big oil still reigns supreme in Alaska and politically speaking it is to the detriment of many Alaskans.

While I have a firm foundation for my low opinion of how big oil controls Alaska's political affairs, it goes even lower when it comes to dealing with issues surrounding climate change. When dealing with big oil there *is no longer a middle of the road.* This was first brought home to me in 2009 when the oil companies walked away from the consensus position of USCAP which formed the basis for the Waxman-Markey cap and trade bill on carbon emissions. Cap and trade is a system of setting up a limit (a cap) on certain types of emissions or pollution, and then permitting companies to sell (trade) the unused portion of their limits to other companies that are struggling to comply. This structure is meant to provide companies with a profit incentive to reduce their pollution levels faster than their peers. The cap and trade approach worked when the US was addressing the issue of acid rain, and as such it was always viewed as a corporate-friendly alternative to the command and control format of most government regulations.

Then before the Waxman-Markey bill could get a hearing in the Senate, the major international oil companies walked away from ever supporting climate legislation and instead embarked on a sophisticated campaign to spread doubt about the science of climate change. The climate denial campaign funded by Exxon[96] and other fossil fuel interests was so successful that it took the US Senate another five years, until January 2015, to acknowledge that "climate change is real and is not a hoax."[97]

As if this wasn't appalling enough, we learned in November of 2015 that the New York attorney general has begun an investigation of Exxon Mobil to determine whether the company lied to the public about the risks of climate change or to investors about how such risks might hurt their investments.[98] The investigation focused on whether statements made by Exxon to investors about climate risks as recently as 2015 were consistent with the company's own long-running scientific research into opening up the Arctic to oil exploration. Apparently, the science was good enough for their operational planning but not for their shareholders.

The campaign of deceit by Exxon is rightly being compared to the tobacco industry's coordinated campaign to deny any cancer link to tobacco; only this time it's the whole Earth that is plagued by a cancer equivalent. Exxon is being accused of funding groups from the 1990s to the mid-2000s to deny climate risks.[99]

Independent from the New York Attorney General investigation, Drexel University recently completed a study aimed at probing the organizational underpinnings and funding behind the climate denial movement. The study found that the largest, most-consistent money fueling the climate denial movement came from a number of well-funded conservative foundations built with so-called "dark money," or money flowing through third-party, pass-through foundations whose funding cannot be traced.[100] The accounting analysis by Professor Robert Brulle revealed that the 'dark money' side of the ledger had been rising dramatically over the past five years while the traceable donations from the

Koch Industries and Exxon Mobile have all but disappeared. In other words, the campaign of deceit still continues; it just went stealth.

Imagine the strides we could have made toward addressing climate change if we didn't have those many years of climate denial subterfuge. Are you not completely and utterly morally outraged by Exxon's push for profit over planet? Exxon's actions are the ultimate example of the indiscriminate and destructive nature of Earth Inc. (to be explained in the next chapter).

In September, 2015, the United Nations climate change panel, the IPCC, warned that at the current emission rates the world will within 30 years exhaust its "carbon budget"—the amount of carbon dioxide it could emit without going into the danger zone above the 2 degree threshold beyond which scientists declare severe drought, rising seas and supercharged storms as well as food and water security become routine challenges.[101] With this carbon budget and timeclock in mind, it's clear that a sizeable amount of fossil fuels will need to stay in the ground. An analysis done by the University of College London's Institute for Sustainable Resources, looked at which regions of the world would need to dial back on development of oil, gas and coal reserves if there is any hope for the global temperature rise to remain under 2 degrees Celsius. [102]

Accounting for regional production costs, size of reserves, production trends and carbon dioxide emissions, researchers Christophe McGlade and Paul Ekins were able develop models of likely fossil fuel exploitations with and without carbon capture sequestration. Their overarching assumption was "that to have at least a 50 per cent chance of keeping warming below 2 °C throughout the twenty-first century, the cumulative carbon emissions between 2011 and 2050 need to be limited to around 1,100 gigatonnes of carbon dioxide (Gt CO_2)." Unfortunately, McGlade and Ekins found that "greenhouse gas emissions contained in present estimates of global fossil fuel reserves are around three times higher than this level. "Our results suggest that, globally, a third of oil reserves, half of gas reserves and over 80 per cent of current coal reserves should remain unused from 2010 to 2050 in order to meet the target of 2 degrees Celsius. We show that development of resources in the Arctic and any increase in unconventional oil production are incommensurate with efforts to limit average global warming to 2 degrees Celsius."[103]

The analysis by the University of College London clearly shows that we are now in extreme times. Even the global insurance companies know this. In September of 2016, leaders of multi-national insurance giants Aviva, Aegon and Amlin, which together manage $1.2 trillion in assets called upon G20 countries to commit to ending coal, oil and gas subsidies within 4 years. "Climate change in particular represents the mother of all risks—to business and to society as a whole, and that risk is magnified by the way in which fossil fuel subsidies distort the energy market," notes Aviva CEO Mark Wilson.[104]

Even without subsidies, the question becomes will oil and gas companies step up and willingly strand their carbon assets to avoid the 2 degrees Celsius threshold? Judging by the sustained campaign of climate denial and the "bait and switch" tactics big oil deployed during the USCAP days, it's safe to suggest that big oil has every intention of pushing the planet beyond the boiling point.

To counter this oil profit or planet be damned scenario, environmental leader Bill McKibben and his group 350.org have sparked a campaign to disinvest University funds from fossil fuel companies. They chose to impact the same bottom line that drives Exxon's actions. Within six months of 350.org's campaign launch, disinvestment efforts were being pursued on over three hundred campuses and in more than one hundred U.S. cities, states and religious institutions. According to Naomi Klein, "no tactic in the climate wars has resonated more powerfully."[105] According to 350.org's website on the ongoing disinvestment campaign, as of November, 2016, there are now 619 institutions actively disinvesting from fossil fuels (some are coal only) and the total value of these disinvesting institutions is approximately $3.4 trillion.[106] The largest institution leading this disinvestment effort is the world's richest sovereign wealth fund, Norway's Government Pension Fund Global worth $850 billion. Even though

Norway's wealth fund is actually founded on the nation's oil and gas wealth they recognize the regulatory risk of investing in carbon intensive companies and as such have divested from 22 companies because of their high carbon emissions: 14 coal mines, five tar sand producers, two cement companies and one coal-based electricity generator.[107]

As Sara Blazevic, a disinvestment organizer at Swarthmore College puts it, the movement is "taking away the hold that the fossil fuel industry has over our political system by making it socially unacceptable and morally unacceptable to be financing fossil fuel extraction."[108]

"Divestment serves as a key tool in moving the world beyond fossil fuels and towards renewable energy," says Payal Parekh, global managing director for 350.org. "The divestment movement is modeling what governments need to be doing: withdrawing funds from the problem and investing in the solution."[109]

Leading financial groups, including Goldman Sachs, Citigroup, and Standard & Poor's, have also warned of the risk posed to fossil fuel investments by action on climate change. In September of 2015, even the Rockefellers, heirs to the fabled Standard Oil fortune, withdrew their funds from fossil fuels. These are encouraging signs that the disinvestment campaign is rising to the challenge.

It should come as no surprise that the disinvestment campaign has caught the attention of big oil and they are pushing back. Bill McKibben is now one of several targets for Richard Berman, a veteran political consultant who has solicited up to $3 million from oil and gas executives to finance his "Big Green Radicals" campaign. "If the oil and gas industry wants to prevent its opponents from slowing its efforts to drill in more places, it must be prepared to employ tactics like digging up embarrassing tidbits about environmentalists," claims Mr. Berman.[110] In a speech to a room full of industry executives that was secretly recorded, Mr. Berman said, "Company executives must be willing to exploit emotions like fear, greed, and anger and turn the public against environmental groups. And major corporations secretly financing such a campaign should not worry about offending the general public because you can either win ugly or lose pretty." In this regard, Mr. Berman paints Bill McKibben as a false prophet whose "prescription for the US would be an economic and social disaster."[111]

If Bill McKibben is being targeted then it is time to win ugly back. The only possible counter to this pattern of subterfuge, deceit, and backstabbing is Blockadia. As captured by the protest stories in Naomi Klein's book, in Blockadia there are no moments of indecision, only place-based stands of action. She reports, "The collective response to the climate crisis is changing from something that primarily takes place in the closed-door policy and lobbying meetings into something alive and unpredictable and very much in the streets and mountains, and farmers' fields, and forests."[112] The climate crisis compounded by the tactics of big oil put us all in extreme times; extreme times that call for the rise of Blockadia.

According to an Economist article (May 2016), "Greens in Pinstriped Suits", Blockadia is even spreading into shareholder meetings. Four of the five largest private oil companies will face shareholder resolutions calling for climate change accountability in one form or another.[113] And when some of these shareholders represent large billion dollar pension funds, board members up for election pay attention.

In her book,_Ms. Klein elaborates further on the protest value of Blockadia:

The power of this ferocious love is what the resource companies and their advocates in government inevitably underestimate, precisely because no amount of money can extinguish it. When what is being fought for is an identity, a culture, a beloved place that people are determined to pass on to their grandchildren, there is nothing companies can offer as a bargaining chip. And though this kind of connection to place is surely strongest in Indigenous communities where the ties to the land go back thousands of years, it is in fact Blockadia's defining feature. It is not the hatred of the coal and oil companies, or anger, but love that will save the place.[114]

There are no bargaining chips in these extreme times created by big oil's denial campaign. For the sake of our planet it's time to tap that deep love of place. With three times more oil and gas in accessible reserves that can be burned if catastrophic climate change is to be avoided it's time to go all out, even if, short of violence, it gets ugly.

Chapter 16 — A Return to Whales

Figure 17: Bubble Feeding Humpback Whales. Kate's photo catches the feeding frenzy associated with bubble netting of herring. Note the lead whale breaking out on top.

As noted earlier, to escape the extreme times and depressing events, to be able to continue the good fight, I return to the respite of Nature. You now know about two humpback whales breaching, but do you know about synchronized bubble-feeding humpbacks? Bubble-feeding occurs when a group of five to twenty-five humpbacks encircle a school of herring. The lead whale dives, followed by another and another. From the surface, I see gigantic flukes (horizontal tails) point toward the sky and disappear in a lovely sequence like an orchestrated water ballet. Underwater, the whales release air bubbles as they spiral down around the herring. Too spooked to swim through the bubble barrier, the herring crowd into a tight swirling ball of fish. The lead whale then gives a vocal signal, a distinctive piercing click. Upon hearing the click, the whales turn in unison to rise rapidly through the netted herring with their accordion like mouths in full extension.

Suddenly on the surface, gulls and terns coalesce from every direction and swoop down toward the water as the ocean erupts with the heads of lunge-feeding whales. In the center, the lead humpback shoots above the other heads, eight feet out of the water, mouth agape. The whales thrash about, lunge-feeding in one of the earth's most impressive predatory frenzies. The high-pierced screams of the gulls accentuate the thrill of capture. Sea lions frolic in the wake of plenty. It is truly a sight of boundless bounty and adaptation.

While other whales also use bubbles as barriers to corral fish, nowhere else in the world do whales execute this high level of organized feeding. When I first moved to Juneau in 1992, the notion of whales bubble-feeding was brand new. To actually see it happening was rare. Now it is common during the summer months, and there are tales of ever-larger groups of whales bubble-feeding in entirely new locations such as Prince William Sound, 450 miles west of Juneau on the far side of the Gulf of Alaska. It has taken only a couple of decades for the first few whales that discovered this technique to pass on their knowledge to other whales.

Through intelligence and cooperation these bubble-feeding whales are charting new paths to survival. We can and must do the same. Find more paths only in far less time.

Time is both on and not on our side. In the "on our side" position we are advancing our connected intellect to new scales of consciousness and understanding. According to theoretical physicist and author Peter Russell, "in one year we experience more innovations than the Pharaohs did in a century." In his book, *Waking up in Time*, Russell concludes, "New discoveries and new technologies will lead to further new abilities and new ways of changing the world. Creativity will continue to breed creativity."[115] In other words, creativity is an evolutionary force and will this force evolve in time?

In the "not on our side" position of time, all one need to do to get a sense of urgency is try to recall the last time you read the headline, "hottest year on record". For the last twenty years, it's been a weather headline that appears regularly every year. Indeed the clock of creativity is up against the clock of climate. Which will win out?

As Joni Mitchell sings "we can only look back from where we came and go round and round in the circle game" and when we look back from where we came in terms of humpback whales off the coast of Alaska, we are making progress. Worldwide populations of humpback whales were decimated due to commercial whaling. Currently, whaling is strictly regulated by the International Whaling Commission, with only small numbers being allowed to be taken for aboriginal subsistence purposes. Since commercial whaling was banned, the population of humpback whales has rebounded.

Before whaling began, approximately 15,000 humpbacks are estimated to have existed in the North Pacific. Current population estimates for the North Pacific stock range from 20,000 individual animals, and 10,000 individuals in the Central North Pacific stock. An increase in numbers on the Hawaiian wintering grounds suggests that at least this portion of the North Pacific stock is growing by approximately seven percent per year.[116] With this type of population response NOAA is now proposing delisting these stocks off Alaska from the endangered species list.

If the humpback whale population can come back as a more highly organized, bubble-feeding species, can we not come back as better humans? Such is my hope for the millennial generation; that they will read the signs for being better humans better than my generation has done.

I agree with Gus Speth in that we—mostly my generation—have not been doing a good job on reading the signs on how to be better humans and our irrevocable decision points are fast approaching. The rapids are upon us. In his book, *The Bridge at the Edge of the World*, Speth teaches us to read the signs now.

In our journey down the path between two worlds, we are fast approaching a place where the paths fork. We got to this fork through a long history dominated by two great and related struggles—the struggle against scarcity and the struggle to subdue nature. To win in these struggles we created a powerful technology and forged an organization of economy and society to deploy that technology extensively, rapidly and, if need be, ruthlessly. So successful were these systems and accomplishments that we were swept up in them, mesmerized by them, captivated and even addicted. We thus continued ahead—ever-grander, ever-richer, ever-larger doing what once made sense but no longer did. There were warning signs along the way, we paid them no heed. These signs said things like this:

being, not having
giving, not getting
needs, not wants
better, not richer
community, not individual
other, not self
connected, not separate
ecology, not economy
part of nature, not apart from nature
dependent, not transcendent
tomorrow, not today[117]

Now we must pay attention to these warning signs. Now we must pay attention to bubble-feeding whales showing us the way.

PART THREE   COMING FULL CIRCLE

Chapter 17 — Earth Inc. Meets Happy Planet Index

Figure 18: "Sustainability – It's in Our Hands", 1992. Pen and Ink with Digital Color by Terry Pyles, © Ray Troll/NOAA, courtesy of Ray Troll, Ketchikan, AK.

We all have moments in our lives that shake up our beliefs in the fundamental institutions that govern our society. Raised as an Irish Catholic who once aspired to becoming a nun, my first crisis of faith moment came when the parish pastor who lived next door to my grandparents decided to tear down my grandfather's grand Victorian style house and turn it into a parking lot. In the event that neither of my grandfather's surviving children could move into the house, the house was to be deeded over to the parish for a community center. Although my family was horrified once they learned of the church's intent to demolish the home site for a parking lot, my family had no legal recourse.

Resigned to this disturbing development, my family asked that all members of our family residing in the immediate area be informed of the wrecking day so that arrangements could be made to be out of town. The pastor, who regularly ate Sunday dinner with my grandparents, did not even do this and unfortunately my Aunt Helen happened to be driving by when the first wrecking ball struck, shattering a home of six decades of loving memories. My aunt collapsed. Subsequently, my faith in the institution of the Catholic Church began to crumble. And to this day, I still feel the sense of betrayal whenever Joni Mitchell's song, "Big Yellow Taxi", comes on the airwaves and she sings, "Paved paradise, put up a parking lot".

For many millennials coming into their professional lives, the Wall Street crash of 2008, served as their "wait-a-minute" moment about the institution of American capitalism. The collapse was devastating for many Americans, but millennials in particular felt the sting through high unemployment. In 2010, youth unemployment (sixteen to twenty-four) rose to eighteen and half percent, the highest youth unemployment since 1948.[118] Up sprung the Occupy Wall Street Movement and here too millennials struggled.

Millennial David D. Burstein notes in his book, *Fast Future*, that the millennials who were part of the Occupy Wall Street Movement were frustrated and trying to take meaningful action, "not at business in general but quite specifically at banks and financial companies and the systemic ways our society has come to favor Wall Street and the wealthy at the expense of supporting young people and the middle class."[119]

Now, fast forward to 2016 and according to a poll conducted by the Harvard Institute of Politics, only nineteen percent of Americans ages eighteen to twenty-nine identified themselves as "capitalist" and only forty-two percent said they "supported capitalism".[120] Instead they support "social justice activism" (forty-eight percent) and "progressivism" (forty-four percent). It appears that among millennials there is a crisis of faith in capitalism.

This doubt about capitalism is not confined to America; it extends to Earth Inc. The term "Earth Inc." was first coined by R. Buckminster Fuller in 1973 to capture the total economic system of spaceship Earth. Since then it has become a metaphor for the interconnected global marketplace—the good, the bad, and the ugly. In his book, *The Future,* Al Gore defines Earth Inc. in the context of the digitally-connected world of today. To him, Earth Inc. means "an integrated, holistic entity with a global relationship to capital, labor, consumer markets, and national governments."[121]

According to Gore in his assessment of six drivers of global change, the technological trend of digitizing the workplace and replacing manual labor ends up accelerating the global wealth gap.

> *A positive feedback loop has emerged between Earth Inc.'s increasing integration on the one hand and the progressive introduction of interconnected intelligent machines on the other. In other words, both of these trends—increased robosourcing (replacing workers with robots) of jobs and the interconnectedness of the global economy driven by trade and investment—reinforce one another. As this shift in the relative value of technology to labor continues to accelerate, so too will the levels of inequality.[122]*

Being ever data-driven in his analysis, Al Gore points out that the measurement of inequality in the United States—the Gini coefficient which measures the inequality of income nation by nation on a scale from zero to one hundred, has risen from a low of thirty-five in 1967 to high of forty-five in 2010. And it still is rising. China, Russia, Egypt, and even Tunisia are lower on the inequality scale. According to Gore, "In the United States fifty percent of all capital gains income goes to the top one thousandth of one percent."[123]

Take half-a-minute and think about this: The wealthiest 300 Americans control fifty percent of all the capital wealth generated by the U.S. economy, while the other fifty percent is split among the rest of us *300 million* Americans. This is a travesty of imbalance that has the founders of our constitution turning in their graves.

On a global level, the travesty of imbalance is even more acute. According to a January 2015 analysis by Oxfam, a globally recognized anti-poverty organization, eighty people hold the same combined wealth as the world's 3.6 billion poorest people. They found that "since 2009, the wealth of those 80 richest has doubled in nominal terms while the wealth of the poorest 50 percent of the world's population has fallen."[124]

Gore is not alone in recognizing the limitations of Earth Inc. At a 2012 gathering of the financial elite in Davos, Switzerland, Bloomberg conducted a poll on capitalism. Seventy percent of the attending business leaders believed capitalism to be "in trouble". One-third of these business leaders also saw the need for a "radical reworking of the rules and regulations".[125]

In a feature article Time Magazine's economic columnist, Rana Foroohar, discussed capitalism's great crisis; noting that most of the world's market economies are grappling with aspects of the same disease affecting capitalism.[126] "Globally, free-market capitalism is coming under fire," notes Foroohar, "as countries across Europe question its merit and emerging markets like Brazil, China and Singapore run their own form of state-directed capitalism." Ms. Foroohar like Mr. Gore cites a broad range of financiers and elite business managers that now speak out about the need for a new more inclusive type of capitalism, one that also helps businesses make better long-term decisions rather than focusing only on the next quarter.

Even Pope Francis has become a vocal critic of modern capitalism, criticizing the "idolatry of money and the dictatorship of an impersonal economy" in which "man is reduced to one of his needs alone: consumption."[127] At last, I now agree with the head of the Catholic Church and recognize that Earth Inc. needs an overhaul to reverse this growing cancer of inequality.

To counter this deleterious effect of Earth Inc., Al Gore suggests that the Global Mind must step in and be the counter force. Connecting the wisdom of compassionate thinkers like Gus Speth—being not having, giving not getting, better, not richer etc.—through the internet we can create centers of influence that are not controlled by the one-percent of wealthiest individuals. This is the power of the Global Mind. Gore says:

> *The Global Mind is the simultaneous deployment of the Internet and the ubiquitous computing power creating a planet-wide extension of the human nervous system that transmits information, thoughts, and feelings to and from billions of people at the speed of light. The emergence of the Global Mind presents us with an opportunity to strengthen reason-based decision making.*
>
> *The outcome of the struggle to shape humanity's future that is now beginning will be determined by a contest between the Global Mind and Earth Inc. In a million theaters of battle. The reform of rules and incentives in markets, political systems, institutions, and societies will succeed or fail depending upon how quickly individuals and groups committed to a sustainable future gain sufficient strength, skill and resolve by connecting with one another to express and achieve their hopes and dreams for a better world.[128]*

Where is your theater of battle? My theater of battle for the Global Mind has shifted. No longer is it in legislative committee hearings or in the meeting halls of fishermen. Today my theater of battle lies with writing. Is your theater in personal choices of daily living or as a volunteer in your community? Is it on a bigger stage? Is it a platform like the Internet or social media?

Perhaps one way to begin countering the cancerous side-effects of Earth Inc. is to introduce "happiness indices" into the discussion of economic growth. If you Google up "Happiness Indices", at the top of the web site list is the Happy Planet Index (HPI). Recognizing the need to have economic measurements as if people and a healthy planet mattered, Nic Marks, founder of Happiness Works, created the HPI. The HPI ranks countries on how many long and happy lives they produce per unit of environmental input.[129]

$$\text{Happy Planet Index} \approx \frac{\text{Experienced well-being} \times \text{Life expectancy}}{\text{Ecological footprint}}$$

Unless you've traveled to Latin America you might be surprised to learn that developing countries score the highest in delivering fairly long and happy lives with a relatively low ecological footprint. Costa Rica was the number one country on the Happy Planet Index for 2014. The United States was 105th.

I know that the Happy Planet Index sounds like it comes from Sesame Street, but the philosophy behind it is sound. It is one more way of getting away from equating growth only to levels of production and consumption. We need to transition to a more holistic view of growth, one that takes in well-being measurements in promoting economic growth, where the metric that guides our decisions is not money but happiness. Although this may sound far-fetched to Americans, it's not to the United Nations.

In April 2012, the United Nations, inspired by the government of Bhutan (first country to measure and use happiness indices) held a "High Level Meeting for Wellbeing and Happiness: Defining a New Economic Paradigm" with over 650 civic leaders from around the world.[130] The meeting marked the launch of a global movement to shift away from measuring and promoting just economic growth as a goal in its own right toward integrating human happiness and quality of life into the equation of economic growth.

Making a policy connection between resident happiness and policy is one of the many important takeaways from the U.N.'s latest World Happiness Report. Here are some of the other noteworthy conclusions:

- Despite the obvious happiness impacts of the financial crisis of 2007-08, the world has become a slightly happier and more generous place over the past five years.
- There has also been some progress toward equality in the regional distribution of well-being.
- Levels of subjective well-being are found to predict future health, mortality, productivity, and income, controlling statistically for other possible determinants.
- There is now a rising worldwide demand that policy be more closely aligned with what really matters to people as they themselves characterize their lives.[131]

Have you noticed that when we switch our analytical focus from economic well-being and equality to happiness trends, the outlook on the world improves? Remember those seventy percent of world business leaders that think capitalism is in trouble? Could this be because economic policy is a not closely aligned with what matters most to people? Think about what a giant step it would be if we paid as much attention to raising the United States happiness ranking as we do to raising the GDP. By the way, even after accounting for the GDP disparity Costa Rica is still happier than the United States.

The U.N.'s report is valuable in other ways than just comparing happiness rankings. For example, the report examines how our collective view of happiness has changed over time.[132] Historically speaking, it is relatively recent that society began associating happiness with economic well-being. If we return to a more holistic perception of happiness we get closer to the ancient insights of Buddha and Aristotle. The report suggests that it is time to replace the "utility theory" of happiness with a more robust outlook that brings us closer to past virtues.

According to the U.N.'s historical analysis, we can raise "evaluative" happiness in society by a return to the virtues taught by Buddha, Aristotle,

Mohammad, and Jesus. As such the report advocates a return to "virtue ethics" as one way to raise happiness in society. In Alaska, this would include a return to the virtues taught by Native elders. Tlingit people (the primary native group in my community) believe that all life is of equal value; plants, trees, birds, fish, animals, and human beings are all equally respected. Although I am not Tlingit, I believe the same.

In the struggle between Earth Inc. and the Global Mind, we must not forget about the ability of one to transform the other. We can change how Earth Inc. operates. Instead of nurturing business as usual, we can adopt new business models that avoid the disturbing feedback loop that Al Gore warns about and aligns better with what matters most to people. It will bring us closer to Gus Speth's virtue of plenitude.

In his book, Speth, advocates ten "virtues of necessity". One of his virtues is the virtue of plenitude.

> *Plenitude. Consumerism, where people find meaning and acceptance through what they consume will be supplanted by the search for abundance in things that truly matter and that bring happiness and joy-family, friend, the natural world, meaningful work. Individuals and communities will enjoy a strong rebirth of reskilling, crafts, and self-provisioning. Overconsumption will be replaced by new investment in civic culture, natural amenities, ecological restoration, and education and community development.*[133]

Well before the time when the concepts of sustainability were well understood and articulated, environmental author, Edward Abbey wrote, "Growth for the sake of growth is the ideology of cancer." Was Abbey prescient about consumerism and over consumption? Perhaps. Nonetheless, I would like to change out the cancer cells metaphor with stem cells. If growth for the sake of growth is the ideology of cancer, then growth for the sake of sustainable communities is the ideology of stem cells.

Switching economic metaphors about growth makes even more sense when you consider how little wealth by US standards it takes to generate happiness. A 2010 Princeton University study by Daniel Kahneman and Angus Deaton found that a single person making more than $75,000 per year will not significantly improve their day-to-day happiness by making more money—$75,000 is the break-off point for happiness.[134] This suggests that the US economy could make a U-turn on the inequality trend without lessening the overall happiness of the country. In other words, The Happy Planet Index could counter balance Earth Inc. Score one for the Global Mind.

Given the clarion voice of Pope Francis to speak out against the misguided idolatry of money around the world, score two for the Catholic Church. He has taught me more than anybody that with reform, faith can be restored.

Chapter 18 — From Baked Alaska to Obama

Figure 19: "SOS on Acid Ocean", 2009. Aerial art by John Quigley, photo by Scott Dickerson, © Scott.Dickerson.com, Homer, AK.

Baked Alaska is not just an ice-cream cake in a mound of toasted meringue, it is the summer I am experiencing. Hardly a year goes by when the weather maps showing current summer temperatures compared with the ten-year average don't show large swatches of Alaska in the red (red being the color for the greatest difference) zone. Off for a weekend camping trip, I was perched in one of those hot red zones—on a beach, in a collapsible camp chair designed for my sleeping pad.

Sitting on the ground, I was able to rock back and forth, taking in either the grand view of the snowcapped Chilkat Mountains framing the far side of Lynn Canal or twisting slightly to take in the calm, forested island view. Sea lions were barking to the south while songbirds were singing in the north. In the background, holding a steady rhythm, waves lapped softly. The whistling wings of scoters took off in flight with the soft yet high-pitched sound of wings beating in unison. I was at peace with Nature.

When nature is at peace we must appreciate it so that we might understand the nature of peace. —Anonymous

In the mid-distance view from where I sat in my snug camp chair, my husband was anchoring out our skiff. He was puzzling over tides and anchor lines. He scratched his beard as he laid out more line, forming a crisp silhouette of thought in action. All around me were calm, reflective waters. It was actually *hot*. I was getting a great tan in Baked Alaska. Climate change never felt or looked as good as it did at that moment. Soaking in all this warmth and beauty of a day, it was tempting to say, "If this is climate change, bring it on!"

But I didn't, because of the salmon waiting for more rain to raise the water levels in the creeks before they could swim upstream, home to spawn. I didn't say "bring it on" because the very waters that were lapping softly in the background of my summer bliss are the ocean waters most at risk to ocean acidification. Colder water absorbs more carbon than warm tropical waters, and the shallow waters of Alaska's continental shelf also retain more carbon dioxide due to less mixing of seawater from deeper ocean waters.

Understanding ocean acidification starts with the fact that the oceans are a carbon sink, taking up to one-quarter of all the carbon dioxide emitted into the atmosphere. Once in the ocean, CO_2 reacts with water to form carbonic acid and carbonate. Carbonate is good as it forms the basis of shell-like structures for a grand diversity of creatures, from zooplankton to Alaska King Crab, whose very survival is dependent on making shells.[135] The problem is that the more acidic the ocean becomes due to absorbing carbon and producing more carbonic acid, the harder it is for the shell-bearing creatures to build their shells. In some cases the water can become acidic enough to break down existing shells. Zooplanktons like pteropods (also known as sea butterflies) are the pastures of the ocean food chain, so the entire marine food web is exposed to the threat of ocean acidification and there is now evidence that this vital part of the web is dissolving.[136] Because of this threat, ocean acidification is often referred to as the evil twin of climate change. And it's already here. Scientists estimate that the ocean is twenty-five percent more acidic today than it was 300 years ago.

No one knows more about ocean acidification in Alaska than Dr. Jeremy Mathis, supervisory oceanographer at NOAA's Pacific Marine Environmental Laboratory and Director of the University of Alaska's Ocean Acidification Research Center. "It seems like everywhere we look in Alaska's coastal oceans, we see signs of increased ocean acidification," warns Dr. Mathis. Due to the threat to pteropods (that salmon and other commercial species eat), the increasing acidification of Alaska waters "could have a destructive effect on all of Alaska's commercial fisheries. This is a problem that we have to think about in terms of the next decade instead of this century," explains Dr. Mathis.[137]

In September 2009, commercial fishermen from Homer joined together with other mariners to send an SOS distress signal (see Figure 19) about ocean acidification. On a day of calm waters and broken clouds, forty-three kayakers joined in formation to spell out "Acid Ocean." Encircling the kayakers in an oblong "O" were an equal number of fishing boats. This bird's eye view of their graphic was widely distributed and promoted as an SOS for the world's oceans.

This message was sent out more than half a decade ago. Since then the signs of acidifying oceans have only increased. Since then President Obama came to Alaska on a mission of ringing the alarm bell about climate change. It was all part of his strategy of setting the stage for the all-important Paris climate negotiations.

In anticipation of the president's climate change visit to Alaska, I set up a meeting with the Lieutenant Governor to discuss the necessity of addressing climate change as more than opening up the Arctic to development. In that meeting, he asked me if I was willing to prepare a climate change briefing paper that summarized impacts occurring in the state and actions by previous administration. He also asked me to focus on recommendations that did not involve legislative approval. I gladly accepted this task.

Wanting to produce a tight, up-to-date paper that touched on the latest developments in renewable energy and Arctic policy, I called upon four other environmental and renewable energy leaders to assist me. Within weeks, we produced a fifteen-page briefing paper that had twenty policy recommendations, including this push to take advantage of the president's upcoming visit:

> On August 31 (2015) when the President of the United States and the Secretary of State—along with foreign ministers from many other countries—gather in Anchorage for a day to highlight climate impacts on the Arctic, Alaska's leaders should present a strong show of support for policies and practices that will benefit the state's people, economy, cultures, wildlife and environment.

Two weeks after submitting the paper and about a month before the President's visit, the five of us who authored the paper met with the Lieutenant Governor. The meeting can only be described as underwhelming. Polite, respectful but no resolve for action; no desire to take advantage of having President Obama here on a mission of climate change.

Not wanting to let all the good work that went into producing this briefing paper go to waste, I called a friend of mine who had recently received a significant appointment and explained the value of the briefing paper. I then sent it along with the knowledge the paper would get discreetly circulated within their circles of influence. The paper was seen as a timely addition to the background papers for the Conference on Global Leadership in the Arctic: Cooperation, Innovation, Engagement and Resilience, called GLACIER for short.

Unable to get an invite to the climate conference in Anchorage, I went camping with my husband. With the boat now anchored, we were both able to relax and enjoy the splendor of a magnificent day in what felt like paradise.

In the heat, the water looked absolutely inviting. I jumped in to cool off. The tide had just washed over the sun-warmed rocks and I could swim without going numb; the upper few feet of water were much warmer than the fifty-four degree water below. Floating on my back, I watched an eagle ride a thermal above me. Ah, yes, I was loving Baked Alaska. But still I did not wish to bring it on.

Just this week came a report of four humpback whales floating dead in the Gulf of Alaska. Scientists suspect that a strong El Nino and warm North Pacific Ocean water contributed to poor feeding conditions for these whales. Warmer water may keep me swimming longer, but the downside for the whales and other marine creatures is tragic.

I am scared most about ocean acidification. With the oceans producing more than fifty percent of the world's oxygen and absorbing a third of the world's carbon, they are the planet's lungs. Yet we continue to expose them to the cancer of indiscriminate growth. We are altering entire ecosystems in ways that we have no possibility of meaningfully reversing. President Obama understood this when he gave his speech before the GLACIER conference in Anchorage.

Climate change is already disrupting our agriculture and ecosystems, our water and food supplies now. If we do nothing, temperatures in Alaska are projected to rise between six and 12 degrees by the end of the century, triggering more melting, more fires, more thawing of the permafrost, a negative feedback loop, a cycle—warming leading to more warming—that we do not want to be part of. And the fact is that climate is changing faster than our efforts to address it. That ladies and gentlemen, must change. We're not acting fast enough. I've come here today, as the leader of the world's largest economy and its second largest emitter to say that the United States recognizes our role in creating this problem and we embrace our responsibility to help solve it.[138]

Although my preference would have been to hear these words of commitment directly at the Anchorage conference, I was feeling renewed by my days of beach camping in Baked Alaska. After dinner Bill and I shared a bottle of wine over the campfire, and our discussion wandered as we wistfully observed the flames. As Louis Pasteur once said, "A bottle of wine contains more philosophy than all the books in the world."

After dousing the campfire, we took a short walk down the cobblestone beach and into the woods. At ten in the evening, there was still plenty of light and in the thick duff of spruce needles, I found a single calypso orchid balanced on a slim stalk. It was a rare find. "How did it get here on this tiny island?" we wondered in appreciation for all the surprises that Nature continually sends our way. Its pink-speckled petals spoke of a delicate beauty amidst unknown change. Its solitary nature spoke of resilience.

Upon my return home, I had a voicemail message from the friend to whom I shared the Climate Briefing Paper with. Apparently the president's advisors had found my paper very useful and had even used it to brief the president. I played the message again and felt the bloom of persistence amidst the duff of disappointment with the State of Alaska.

Chapter 19 — The Great Unconformity

Figure 20: Into the depths and light of floating through the Grand Canyon in winter. Photo taken by Kate, November–December, 2011.

Passing the "spontaneity test" is an important skill to have if you're going to live in a rainforest for thirty-some years. Although Alaska is indeed heating up, the rainforests in Southeast Alaska are predicted to get wetter with more precipitation coming in the form of rain instead of snow. The rarity of bright blue days has trained us Southeasterners to be spontaneous and willing to alter our schedules at the first sign of a partly cloudy day. Regardless, I was still flummoxed when a former rafting companion asked me with just three weeks' notice, "Do you want to go on a twenty-seven-day raft trip through the Grand Canyon?"

Over five million visitors are drawn to the Grand Canyon every year, the most popular National park in the United States. Of these five million visitors, only 25,000 get to experience the solitude and thrill of rafting through all 280 miles of the Colorado River.[139] More than half of these 25,000 rafters are on commercial trips. A little over two percent of the annual visitors to the Grand Canyon experience the park through privately organized raft trips, and even fewer go in the month of November. Knowing the uniqueness and rarity of this opportunity to experience the Grand Canyon in near solitude, I was compelled to do whatever it would take to seize the offer. Adventure was not just calling, it was screaming.

When not in the raft with my trip companions, I rotated with others in and out of an inflatable kayak. I particularly enjoyed the hours of aimlessly floating and twirling about in the many-hued reflections of sandstone walls. I savored my alone time, whether in the kayak or hiking about. When I hiked, I wrote.

Journal Entry November 2014 – Day 7

This stretch of the river is peaceful, allowing me to journey inside myself for a day. Here in the depths of the finest canyon on Earth, I find I am often looking up, awestruck by beauty. I'm reminded of how my soul soared when I walked inside a Gothic Cathedral for the first time... one of those 'imitations of immortality' moments. That's what I feel constantly and it's only Day 7.

Yet, the weight of time as etched out in the canyon can at times be overbearing. The three-thousand foot canyon walls display time in very distinct layers from a few thousand years to 1.6 billion years. These skyscraper tablets of time are exposed at every bend in the river. No place on Earth lays out time like the Grand Canyon. No place else plunges you into deep time like this place.

All in all, it makes me feel pretty insignificant. Humanity is but a speck on this immense landscape of time. These canyon walls have a legacy that lasts for hundreds of millions of years. How long of an impression, if any, will my life have?

How do I sort out this duality—feeling exalted and insignificant at the same time? But do I need to?

For now it feels best just to let my soul soar among the towering buttes bathed in evening light. Soon shadows emerge, revealing additional layers of depth. With the shimmering reflection in the river, I feel myself blending into the rhythms and the architecture of greatness. I am so very privileged to be here.

Back on the river, I found myself spurred on by the bravado of four young oarsmen. I started running bigger rapids in the kayak. In one set of wild rapids along a canyon wall, I was completely submerged but kept my kayak pointing toward the bright yellow raft I could glimpse downriver. Popping through a back curl of water, screaming with adrenalin, I felt the rush of youth. Later these same oarsmen coached me as I climbed through a narrow slot canyon (when you can touch both sides of the canyon walls with outstretched arms) and slithered deep into the veins of the canyon's architecture.

It was the gift of youth that brought me deeper into the canyons than I had ever expected. My gift in return was to pass on geological tidbits. My favorite geological discovery was a thin, obscure geological formation known as the great unconformity—a name that seemed to fit our eclectic group of oddly connected friends of friends.

Geologically speaking, the great unconformity is where the deep brown Tapeats Sandstone, 500 million years old, sits directly upon the Vishnu Schist, the oldest rock in the canyon at 1.6 billion years old. This results in a time gap of 1.1 billion years within the rock layers. The great unconformity can be seen in Blacktail Canyon, a short hike in from the river's edge.[140] Hiking along the gravelly bottom, I was the first to find the light colored Tapeats Sandstone layer. We followed this layer of sandstone around a bend and soon spotted the dark-rose-colored Vishnu Schist layer. By placing one hand on the curvy Tapeats Sandstone and another hand on the Vishnu Schist, we literally spanned 1.1 billion years with our hands.

The existence of modern man probably measures a mere ten feet in this intense landscape of time. But it is our consciousness that gives rise to the measurement of time. Holding a quarter of the age of the Earth in my hands, I now found the Canyon's landscape less intimidating. The conflict of simultaneously feeling immense awe and complete insignificance seemed to evaporate, replaced by a sense of interconnectedness to the world at large.

Bill Plotkin, a cultural psychologist, wilderness guide and author of *Nature and the Human Soul,* says "Your soul is and is of the world, like a whirlpool in a river, a wave in the ocean, or a branch of flame in a fire."[141] He goes on to explain that "the difference between caring for the world and caring for the soul of the world is the difference between shifting from our smaller personal needs to the world's needs."

In a land revealed by time, I have reconnected to the soul of the world and it matters. This is what I felt when alone in the sunlit depths of the Grand Canyon. I believe the soul of the world can still affect the outcome. That's what these walls of eons were telling me.

At the very top of these walls lies a thin layer of loose rock that represents the Anthropocene era. The Anthropocene era is the term that geologists use to represent the imprint of man, all 250,000 years of our collective existence since *Homo sapiens* evolved on planet Earth.[142] When we ponder the imprint of modern mankind as represented by today's 6.5 billion people all seeking a middle class lifestyle—the disruption of habitat, the extinction rate of two hundred species a year, the ability to pollute to a level capable of altering weather patterns and acidifying the world's ocean—we see clearly that mankind is an evolutionary force. Instead of being a species forever evolving through the forces of Nature, we are now the sole species forcing Nature to evolve on our terms. This is the other great unconformity. It too is missing from these walls, not because the river has eroded it away, but because its time span is too short to be noted in these layers.

It is my hope, my belief that the "great *awareness* unconformity" of mankind, the one that can hold a billion years between their hands, the one that is part of the Global Mind, wins out in the end over the "great *disruptive* unconformity," Earth Inc. Let the soul of the world redefine the Anthropocene era in this way. This is my canyon dream.

There were days in the canyon when the sense of time, although pervasive, was not the dominant sense by any means. Fun still prevailed. The water toys—paddle board and hard shell kayak—came out time and time again. Day twenty-three, however, was to be a serious day, as Lava Falls, the most notorious of the big rapids, awaited us.

Still a half-mile away, the roar of the rapids, told us that after 179 miles on the river we were entering new territory. All sixteen of us scurried along the boulders to scout the river. There was no obvious safe channel. True to its reputation, Lava Falls growled before us, a rippled mass of lurking holes, big hydraulics and back curls. I began to feel a bit paralyzed by the size and torment of what laid before me.

Similarly, one can be paralyzed by the magnitude of the climate change challenge. I have often felt the despair that Al Gore warns against. I know I need to think like climate change—broadly across the canyon landscape— and look deeper in the future, but still that sense of dread lingers on the edges. In the *Rolling Stone* article "The Turning Point", Gore explains:

> In the struggle to solve the climate crisis, a powerful, largely unnoticed shift is taking place. The forward journey for human civilization will be difficult and dangerous, but it is now clear that we will ultimately prevail. The only question is how quickly we can accelerate and complete the

transition to a low-carbon civilization. There will be many times in the decades ahead when we will have to take care to guard against despair, lest it become another form of denial, paralyzing action. It is true that we have waited too long to avoid some serious damage to the planetary ecosystem—some of it, unfortunately, irreversible. Yet the truly catastrophic damages that have the potential for ending civilization as we know it can still—almost certainly—be avoided.[143]

Coming out of the temporary paralysis caused by the site of Lava Falls, I had to believe that like climate change with skilled leaders and proper planning flipping a raft in Lava Falls could be avoided. Watching the torrent of cascading water before us we discussed who would be in each raft and each person's individual role.

According to *Rolling Stone* Reporter Jeff Goodell who closely followed the Paris climate negotiations, the success of the Paris agreement comes down to individual actions. "In the end, the most striking thing about the Paris agreement may be the degree to which it bets the future of civilization on individual actions. The summit was not about the establishment of a global carbon police force—it was an attempt to write the rules so that people are inspired to go out and fix the problem on their own, either as individuals or as nations, and to lend a hand to others who are less fortunate. And depending on your view of human nature, that may be the most risky bet of all," concludes Goodell.[144]

My individual role in being in the first raft to enter Lava Falls, was to yell 'forward' when it came time to throw our body momentum toward the bow in order to keep the raft from flipping. I knew my oarsman well and could read his body language of urgency.

A major concern for our group was how to get a transport catamaran through the boiling hydraulics with just one oarsman. It was a craft not of our party. We had inherited it when we assisted a capsized rafter earlier in the trip. The solo rafter had ridden with us to the Park Service's office at Phantom Ranch, where he was able to ride a mule to the rim with his personal gear. The Park Service had asked us to transport his catamaran with us to the take-out point at Lake Mead. We had agreed, despite knowing that the small catamaran had a faulty design of which any raging river could take advantage.

Our companion and experienced oarsman, Eugene, rowed alone in the catamaran, following our line of approach. Buried in waves soaring overhead, I glanced back as we all rushed the bow FORWARD! I caught a glimpse of the catamaran as it flipped in the first set of waves and back curls. Eugene was catapulted into a wild ride through a set of ten-foot-tall haystacks. Then the careening, empty catamaran came tumbling down on top of him, and Eugene disappeared. Unable to slow or stop, we shot downstream, waiting anxiously in the first available eddy where we positioned our raft to catch him.

At last we saw Eugene surface. Now, the catamaran, upright and running fast down the middle of the river was heading straight toward Eugene. We screamed and pointed, "Raft coming at you"! He noticed the catamaran and started swimming to the side. Within seconds, he managed to catch the careening craft without getting hit by the wildly swinging oars. Finally, with great determination, bouncing through another set of haystacks, Eugene managed to fall back in the seat holding onto to one oar. He grabbed the other oar, and signaled that he was "okay" by repeatedly slapping the top of his helmet with his open hand.

Even though I am greatly relieved by Eugene's individual effort to reclaim the catamaran, I know that like all of us entering Lava Falls, we must face the turbulent times ahead with great determination. Hardly a week goes by without the evening news reporting on some weather related disaster or scientists reporting on another carbon benchmark being reached. Similar to Lava Falls, the climate is already locked and loaded for a wild ride. Yet, we cannot become paralyzed by the roar of wildness and uncertainty that lies ahead. There will be times when climate activists will feel submerged by this challenge. My advice is to keep yourself pointed straight. Throw your momentum forward as you enter the daunting waves. A raft of fellow citizen-activists will likely pick you up and you will have the satisfaction that each individual action counts.

In his conclusion to *The Future*, Al Gore, says, "Where our journey takes us next will depend upon what kind of beings we humans choose to be. To put it another way, our decision about the way we choose to live will determine whether the journey takes us or whether we take the journey."[145]

My takeaway from the depths of the canyons is to know that the soul of the world says we take the journey instead of the journey taking us. Will you join me on this journey? This is your spontaneity test.

If you answer yes, "be the future of your planet". This line comes from a world renowned theoretical physicist and cosmologist, Marcel Gleiser. I cannot say it any better: *Be the future of your planet.* This is the chance that the 2016 Paris climate agreement has given us to save our planet.

Chapter 20 — Aurora Afterword

Figure 21: Kate skating in front of the Mendenhall Glacier, twenty years apart. 1994. See Figure 22 (2014).

In your efforts to approach climate change, do so creatively and without boundaries. It might be a call to your university to disinvest in coal. Perhaps you'll join in a public outcry against extreme extraction. Or your contribution might be as simple as explaining to the person next to you in an airport lounge that the excessive weather causing the delay is also attributable to climate change. I remain astonished at how little we connect the dots between how unusual the weather is and what we are doing to the atmosphere. Even the colder winters are related to climate change. For example, I ran into a woman from North Carolina going on about the East Coast snow storms of 2014.

She turned to me. "It makes you wonder what in the world is going on."

"I know what's happening," I replied.

She gave me a dirty look. "Don't tell me Al Gore is the answer."

"No," I answered. "Actually, weather scientists speculate that due to a buildup of greenhouse gases, the jet stream that normally flows west to east is moving in a wavy pattern north toward Alaska, giving us rain instead of snow. Then this wavy jet stream hits the Arctic front and diverts all that cold air eastward." I motioned in the same way the weather lady would. "They say this could be the new normal."

"Really? I hadn't heard that."

By avoiding the terms *Al Gore, global warming,* and *climate change…* all those hot button words that Fox News has taught people to sneer at, she actually got the picture and connected the dots herself. In a minor way my conversation contributed toward raising her consciousness about climate change.

Perhaps the most important place to be the future of your planet is as an individual raising a new consciousness of sustainability. What do I mean by this? Turning to Gus Speth, "A new consciousness of sustainability means a more intellectual process of coming to see the world anew and deeply embracing the emerging ethic of the environment and the old ethic of what it means to love thy neighbor as thyself."[146]

For me, the call to raise a new consciousness of sustainability and to think creatively outside one's normal sphere meant moving beyond the world of resource politics and switching to writing as my primary focus. In addition to writing op-ed columns for my local and statewide newspapers, I have written two feature screenplays that deal with the subject of climate change. One is a political thriller, entitled *Hot Ice, Cold Lies*. The other, *Deadliest Cause*, is an adventure drama dealing with ocean acidification.

Taking the gamble on writing this book, on venturing into screenplays as part of being the future of my planet, did not come easy. As an unknown author, writing screenplays is a huge gamble. The odds of them getting produced are as thin as the Anthropocene layer in the Grand Canyon. With the wave of rejection letters that invariably come with a writing career, it's tempting to think about one more opportunity to play that pivotal policy role.

But now, in the quiet of deep reflection (brought about by writing), I no longer hear the second half of the Denali chant: *the seeds for great accomplishments in my field lie within me.* Writing in Crookshaven, Ireland (April, 1973) nine years before I climbed Denali, I had written in my travel journal these words, which I should have re-read before now.

> *Listen to the river and hear it talk of your heart.*
> *Listen to the sea and hear it talk of life.*
> *Then find the harmony of what you hear and live it in your life.*

I think I did this. No longer hearing a reminder of career disappointment, I now hear and write instead about the United Nation's Earth Charter. Here is my abbreviated version:

> *We stand at a critical moment in Earth's history, a time when humanity must choose its future. We have the knowledge and technology to provide for all and to reduce our impacts on the environment. We are at once citizens of different nations and of one world in which the local and global are linked. The spirit of human solidarity and kinship with life is strengthened when we live with reverence for the gift of life, and*

humility regarding the human place in nature.[147]

The harmony of what I hear is akin to touching the soul of the world. Here's a poem to send you on your way to being the future of your planet:

The Hope of Aurora

Shaking off the remnants of dreams, I rise to go outside
The dreams of the earth dance above a stage of white capped mountains
Roving swirls of green light curtains capture stars in their sweeping motion
Streaming golden lights of Juneau reflect in deep waters below
Hope lies where the cosmic dance of aurora meets the gentle light of harmonious man.
I watch in silence, I am renewed in a sense of greater awe.
The magic of wondrous night rolls away into the amber light of dawn.
What happens when instead of it being darkest before the dawn, it is the lightest, the most colorful? Is there a greater order of hope for humanity?
There must be because the beauty I behold is not for me but for the soul of world.

This is my good fortune to know and share. Sustainability and climate change will not be resolved in my remaining lifetime and I want to leave behind three things for the millennial generation, our next great civic generation. First is an apology for not leaving this earth better than I found it; second, I want to leave behind a sustainable path forward in this very complex world of ours, and lastly, I want to leave behind hope; hope grounded in reality and in the continuing inspiration of Nature.

In my search for reality based hope about climate change I've relied a lot on Al Gore, the man who gave us *An Inconvenient Truth* in 2006. As a big picture thinker and global investor in sustainability, he tracks climate change inside out. If he can be optimistic, then so can I.

In his TED 2016 Talk "*The Case for Optimism on Climate Change*", Gore highlights how throughout the world, we are exceeding the best projections for renewable energy. He notes in the year 2000 that the best projections in the world were that by 2010, the world would be able to install thirty gigawatts of wind capacity and that we beat that mark by fourteen times. For solar, the progress is even more remarkable. According to Gore, "the best projections [at the outset of the century] were that we would install one gigawatt per year by 2010 and when 2010 came around, we beat that mark by seventeen times over. Last year we beat it by fifty-eight times over."[148] With this type of exponential progress, Gore feels we're going to win the climate challenge.

Physicist Peter Russell reminds us that creativity itself is an evolutionary force. Man's disruption to planetary ecological functions is the other evolutionary force at play here in the climate calamity challenge. Nonetheless, add in the speed of the Internet, the political engagement of millennials, the rise of Blockadia, the significance of renewable energy having grid parity and I'm betting on the creative force catching up to deter the worst of calamity. In any event, it's time for me to confidently skate in my iceberg free lake at the base of Mendenhall Glacier.

Figure 22: Kate skating in front of the Mendenhall Glacier, twenty years apart. 2014. See Figure 21 (1994).

APPENDIX

Appendix A — Top 20 Environmentally Performing Countries Using Yale's Environmental Performance Index

Rank	Country	Score	Rank	Country	Score
1	Switzerland	87.67	11	Netherlands	77.75
2	Luxembourg	83.29	12	United Kingdom	77.35
3	Australia	82.40	13	Denmark	76.92
4	Singapore	81.78	14	Iceland	76.50
5	Czech Republic	81.47	15	Slovenia	76.43
6	Germany	80.47	16	New Zealand	76.41
7	Spain	79.79	17	Portugal	75.80
8	Austria	78.32	18	Finland	75.72
9	Sweden	78.09	19	Ireland	74.67
10	Norway	78.09	20	Estonia	74.66

Appendix B — Principles of Eco-nomics

Eco-nomic Principle 1: Treat Conservation as a Universal Value

We see *equality, justice, freedom* and *prosperity* as core American values. While we may disagree as to the correct application of these values in various circumstances, we all tend to agree that these are shared values among Americans. *Environmental conservation* should be on this list of shared values. It is time to recognize this after a hundred and fifty years of inspirational writing from Henry David Thoreau to John Muir to Barry Lopez, after half a century of national conservation leadership from Theodore Roosevelt to Al Gore, and after the forty years of living with the Clean Air Act, conservation is here to stay. We all want a job and the ability to raise a family in a clean and healthy environment. Start from this premise when entering any resource use conflict. Assume that the logger cares, too.

Eco-nomic Principle 2: Invest in "Ecologically Tuned-In" Vested Interests

All renewable resource industries have an immediate vested interest in the sustainability of the raw resource. Some industries act responsibly in protecting their vested interest, others do not. This is likely due to short-term perspectives or lack of vision. To an industry, acting responsibly should include recognizing their ecological roots and being engaged in promoting appropriate environmental protection. Seek out the part of industry that is looking after the "seed corn" for everyone else. When the ecological and economic connection is made, vested interests, including those interests created by the privatization of public access, can work to protect as well as to promote their economic interest.

Eco-nomic Principle 3: Provide a Forum to Advocate Balance

The existing public process of "release a document, hold an official public hearing, make a decision", does not provide a forum for seeking balanced solutions or politically achievable compromises. Work sessions and informal conferences should be favored over the 5-minute "state your name for the record" drill. Public hearings are suitable for advocacy but not for conflict resolution. This is, in part, because any political advocate worth their salary will loyally spend their five minutes advocating their organization or business interest and not waste time suggesting a compromise. Whenever possible convene roundtable discussions with a neutral facilitator. Give people a chance to talk as problem solvers.

Eco-nomic Principle 4: Compatibility Works

Dr. Eugene P. Odum, known as the "father of modern ecology", taught us that species diversity is directly correlated with ecosystem stability. Like species sharing the same habitat, industries depend on healthy ecosystems. Similarly, communities with diverse resource-dependent industries are often more economically stable. Because the ecological base of one resource industry so often overlaps into the base of another industry, seeking compatibility among these resource industries can help promote better management of the shared resources. Working with ecologically tuned in industries (discussed in Principle 2) makes a big difference in finding a path to better overall management of the resource base.

Eco-nomic Principle 5: Maximize the Buy-In to Minimize the Conflict

The more that the most affected interests have their fingerprints on a proposed project, action or policy, the less likely those interests will seek the lawsuit route. The more buy-in on the front end, whether procedural or substantive (preferred), the more likely there will be buy-in on the back end. This is a simple rule that *helps* more stakeholders come to the same preferred alternative. I emphasize that this *helps*, although often deeply felt, complicated conflicts need formal mediation and/or collaboration forums to get to 'yes'.

Eco-nomic Principle 6: Seek Local Solutions Based on Respect

Progress in resolving conflict occurs when there is basic respect for the rights and welfare of fellow human beings. The more that individuals in the conflict see each other as neighbors sharing the same economic and ecological community, the easier it is to let respect enter the dispute and become the glue that holds any agreement together. Recognize that folks are most likely to appreciate that none of us are as smart as all of us… go local when given the chance. Sometimes it can be viable to resolve conflict further up the line at the state and national level. However, it is most important to seek a public venue where stakeholders share and show respect.

Eco-nomic Principle 7: Change is a Given

Change is inherent to both ecological and economic systems. Sometimes change offers opportunities to fuse systems, to combine changes beneficial to both Man and Nature. This is a desirable and sometimes an achievable end, particularly in those resource situations/issues where the hands-on approach to restoring environmental health is the only option. Even in resource areas where the hands-off approach is an option, change is a given. Climate change challenges ecosystem management and sustainable development in ways we never imagined.

Eco-nomic Principle 8: Synergy Rules

In a fully evolved ecosystem, waste and pollution do not exist. What is waste to one species is often food for another. In a fully evolved economic system, one company's waste is another company's raw product material. Companies paying attention to the green line—recycling, green marketing, waste reduction, and energy efficiency—are more stable and profitable in the long run than those that ignore environmental practices. Proper use of the marketplace can accelerate the benefits back to the forest, mountains and oceans. Man, at the scale of 7 billion people, and Nature are forever entwined. We all must recognize that progress and prosperity for humankind can, if done wisely and compassionately, be good for the environment as well.

Eco-nomic Principle 9: Take the Long View Home

Define the resource issue in as global a context as possible. Next, define a time horizon long enough for there to be accountability; long enough for the economic benefits of alternative design to be accounted for in labor and economic statistics. Think as far down the line as the dynamics of the issue will allow. The longer the time horizon, the larger the window for finding and promoting synergy. If given a chance to define the context of the debate, cross over toward the middle and take the long view home.

Appendix C — Al Gore's Twenty-Four Reasons for Optimism

A day before climate marches in 162 countries to be capped off by hundreds of thousands marching in the People's Climate March in New York City, Al Gore did a twenty-four hour post countdown for the Climate Reality Project. Here are Al Gore's twenty-four reasons to be hopeful for a fossil free future.

- Renewable energy projects are growing in use and getting cheaper all the time.
- The cost of rooftop solar is now competitive with utility rates in many places.
- Remarkable progress is being made in energy storage.
- The electric grid is evolving—getting smaller and more flexible.
- The electric vehicle industry and market are booming.
- Transportation is getting more efficient and public transit is growing.
- Energy efficiency is improving globally and saving people money.
- Coal fired power plant regulations are now happening in the U.S.; signaling closing the door on fossil fuels.
- Many large business and global brands are going green.
- Faith communities are embracing renewables.
- Youth and students have driven the expansion of clean energy.
- The tide is turning on public opinion.
- Renewables are reducing global poverty and expanding energy access.
- Clean energy saves lives and makes the world more secure.
- Public health is improved by using clean energy.
- Forest capital is being better protected and deforestation reduced.
- Climate smart agriculture is growing and securing livelihoods.
- Many countries are on the right path to meeting emissions and renewable targets.

- China is making huge investments in clean energy.
- Key financial institutions are realizing dirty energy is a bad investment.
- Global, regional, state and local entities are making strides on renewable energy and efficiency
- Cities everywhere are using renewables and energy efficiency for smart growth.
- U.S. Secretary General Ban-Ki-Moon is convening nations toward a strong climate treaty.

Appendix D — "Corporate interests 4, Alaska Zero. Game over?"

(Published July 19, 2013 in the *Alaska Dispatch News*)

As a pivotal global leader, President Franklin D. Roosevelt influenced the growth of democracy around the world. He astutely noted that: "The liberty of a democracy is not safe if the people tolerate the growth of private power to a point where it becomes stronger than their democratic state itself. That, in its essence, is fascism—ownership of government by an individual, by a group." While 'fascism' may seem to be too strong of a word to describe the control of private power in the governing of our state affairs, the warning that Roosevelt issued is all too relevant to today.

The other term that may fit better is Corporatocracy. Corporatocracy is when both the prevailing economic and political systems used to govern a country are controlled by corporations and all their associated interests. Gus Speth, former chair of the United Nations Development Group defines corporatocracy as when corporations move beyond being the principal economic actor into also being the principal political actor as well. In essence, corporatocracy occurs when there is a pattern of having government systems controlled by corporate interest.

Is this where we are in Alaska? Perhaps. Likely, if anyone is keeping score and that would be me. Here is my scoring of recent political events.

Coastal Management

In August, 2012, the coastal management initiative failed at the ballot box. Without the coastal management program, Alaska's coastal communities no longer have a seat at the state permitting table. This is the outcome of having $1.5 million spent by the opponents—Royal Dutch Shell, Exxon Mobil Corporation, ConocoPhillips, BP, Alaska Oil and Gas Association, Pebble Partnership, the Resource Development Council— compared with about $200,000 raised by supporters of the ballot proposition. These numbers come from reports filed with the Alaska Public Offices Commission.

Outside Corporations - 1 Alaska's Citizens - 0

Bipartisan Senate Group Broken

Next came the November 2012 elections when breaking up the Senate Bipartisan Working Group was the main objective of those frustrated with not passing Governor Parnell's $2 billion a year tax giveaway. Major financial contributions from big oil combined with districts re-drawn to favor Republican challengers led to the breakup of the Senate Bipartisan Working Group.

Outside Corporations – 2 Alaska's Citizens – 0

Cruise Ship Waste Roll-back

Then in early February 2013, the Republican-controlled legislature quickly approved a law proposed by Governor Parnell to abolish cruise ship wastewater standards enacted in 2006 by a citizens' initiative. According to the Anchorage Daily News, the new law will allow the cruise industry to indefinitely discharge ammonia, a product of human waste, and heavy metals, dissolved from ship plumbing. Those discharges would have been banned in 2015 under the 2006 citizen initiative. Jim Walker, a maritime lawyer and contributor to Cruise Law News notes, "Formerly the most progressive state in the U.S. protecting its waters from harmful cruise ship discharges, Alaska was intimidated by the cruise industry to roll back its environmental regulations to permit cruise lines to dump high levels of waste by-products and heavy metals like zinc, copper and nickel."

Outside Corporations - 3 Alaska's Citizens – 0

Massive Oil Industry Tax Cut

Now we have Senate Bill 21, the oil tax. The governor's own cost estimate for the bill's potential impact to Alaska's budget is $4.5 billion in just the first five years. The biggest provision in the bill is an across-the-board reduction in Alaska's sale price for our oil. It applies to every drop of oil leaving the state and has no-strings-attached that would require the companies to invest more in Alaska, produce more oil, or hire more Alaskans. Notwithstanding the constitutional directive to use Alaska's natural resources "for the maximum benefit of its people," the Legislature passed SB 21 with two of the key votes in the Senate coming from employees of ConocoPhillips.

Outside Corporations – 4 Alaska's Citizens – 0

Not the Final Score

With a score of 4 to 0, are we at the point that former President Roosevelt warned us about ... the point where the growth of private power is stronger than the democratic state itself? This is scary to think about.

However, the score is not yet final. Citizens and community organizers from across the state turned in over 50,000 signatures to put the question of repealing SB 21 on the August, 2014, ballot. Considering that the organizers only had 90 days to collect all these signatures, this is a most significant show of citizen activism and deserves to be scored. However, the real outcome occurs with the vote. As such, the score stood at Outside Corporation 4, Alaska's Citizens 1/2.

We are a state dependent upon extractive resources, but this does not give these industries the right to extract control of our democracy. With the referendum of repealing the oil tax giveaway we pushed back for a return to the owner state.

1   Sylva Earle, "How to protect the oceans," *Ted Talks,* February, 2009: https://www.mission blue.org/hope-spots/

2   John B. Cobb, "Ten Ideas for Saving the Planet," *Jesus, Jazz and Buddhism.* January 2012.

3   Douglas Fischer, "Dark Money Funds Climate Denial Effort," *Scientific American*, December 2013, https://www.scientificamerican.com/article/dark-money-funds-climate

4   Naomi Klein. *This Changes Everything* (New York: Simon and Schuster, 2014), 226-228

5   Kate Troll, "The line not blurred by Parnell – oil and gas development at all costs," *Juneau Empire* (Juneau, Alaska) August 14, 2011

6   Mark Kelley, "Mendenhall Glacier, Juneau, Alaska – Image 2841," *Mark Kelley Photography* (Juneau, Alaska) Summer 2013

7   NOAA's National Climatic Data Center, "State of the Climate Reports," December 2012

8   Doyle Rice and Alia E. Dastagir, "One year after Sandy, 9 devastating facts," *USA Today* (McLean, Virginia), Oct. 29, 2013

9   NOAA's National Centers for Environmental Information, "Billion Dollar Weather and Climate Disasters: Table of Events" U.S. Drought and Heatwave 2012.

10   United Nation's Conference, "World People's Conference on Climate and the Right of Mother Earth," *Links – International Journal of Socialist Renewal.* Conference held in Cochabamba, Bolivia, April 2010.

11   Jeff Spross, "At this rate, the world will have to cease all carbon emissions in 2040 to stay under 2 degrees Celsius", *Climate Progress,* posted November 17, 2014

12   White House Office of Press Secretary, "Fact Sheet: White House Launches American Business Act on Climate Change," *White House Briefing Room.* July 27 2015. https://www.whitehouse.gov/the-press-office/2015/07/27/fact-sheet-white-house-launches-american-business-act-climate-pledge

13   Morley Winograd and Michael Hais, *Millennial Momentum: How a New Generation is Remaking America* (New Brunswick, New Jersey: Rutgers University Press 2011), 15-18

14   Ibid., 233

15   William Wordsworth, *Wordsworth and Coleridge: Lyrical Ballads* (London and Glasgow: Collins Publishers 1968) p. 138-142.

16   Yale Center for Environmental Law and Policy, "Global Metrics for the Environment," *Environmental Performance Index* (New Haven, CT: Yale University 2016), 110-111

17   Justin Gillis "A Tricky Transition from Fossil Fuels: Denmark Aims for 100 Percent Renewable Energy", *The New York Times*, November 11, 2014.

18   David Burstein. *Fast Furious: How the Millennial Generation is Shaping our World.* (Boston, Mass. Beacon Press Books 2013), 14-19

19   Lynn O'Shaughnessy, "20 most popular study abroad destinations," *Moneywatch – CBS News,* November 2013. www.cbsnews.com/news/20-most-popular-study-abroad-destinations/

20   Thomas Friedman, *Hot, Flat and Crowded: Why we need a green revolution and how it can renew America,* (New York City, New York: Farrar Straus Giroux 2008) p. 177

21   Ibid, p. 177

22   Kate Troll. *Eco-nomics and Eagles.* (USA: Xlibris Corporation, 2002), p. 13

23   Ibid, p.14

24   Ibid, p. 104

25   Chief Seattle. *Wikipedia.* https://en.wikipedia.org/wiki/Chief_Seattle

26   White House Office of Press Secretary, "President Obama Protects Alaska's Bristol Bay from future oil and gas drilling," *White House Briefing Room,* December 2014. https://www.whitehouse.gov/the-press-office/2014/12/16/president-obama-protects-alaska-s-bristol-bay-future-oil-and-gas-drilling

27   Naomi Klein. *This Changes Everything.* (New York: Simon and Schuster, 2014), 438 – 446.

28   Ibid, 444.

29   United Nations "World People's Conference on Climate Change and the Rights of Mother Earth," Cochabamba, Bolivia. April 2010. https://pwccc.wordpress.com/support/

30   Naomi Klein. *This Changes Everything.* (New York: Simon and Schuster, 2014), 444.

31   Kira Gould and Lance Hosey, *Women in Green: Voices of Sustainable Design* (Seattle, Washington: Ecotone Publishing, 2006) p. vi - vii

32   Hanna Rosin, "New data on the rise of women," *TED talks,* December 2010

[33]   Sally Helgesen, *The Female Advantage: Women's Way of Leadership,"* (New York City, New York: Doubleday Publishing April 1990)

[34]   Milton Friedman. *Wikipedia.* https://en.wikiquote.org/wiki/Milton_Friedman

[35]   Catalyst and Harvard Business School. "New Catalyst Study Links More Women Leaders with Greater Corporate Social Responsibility," *Catalyst: Changing Workplaces, Changing Lives.* March 2008.

[37]   Ray Anderson, "Women's Network for a Sustainable Future - WNSF 7th Annual Award to Ray Anderson," *The Glass Hammer Network,* 2005.

[38]   Jackie Vanderburg, "The Global Rise of Female Entrepreneurs," *Harvard Business Review*, September 4 2013

[39]   Ibid, p. 1

[40]   Marge Piercy, "Circles on the Water" *The Hunger Moon: New and Selected Poems, 1980 – 2010,* (New York City, New York. Knopf Publishing 2011)

[41]   Chipko movement, "Women and the Environment," *Wikipedia* https://en.wikipedia.org/wiki/Women_and_the_environment#Ecological_movements_initiated_by_women

[42]   Jeff Goodell, "Interview with Gus Speth: Change Everything Now," *Orion Magazine,* September/October 2008. p. 1

[43]   Ibid., p. 5

[44]   Ibid., p. 3

[45]   Marine Stewardship Council, "Certified Sustainable Seafood," https://www.msc.org/get-certified/fisheries/eligible-fisheries

[46]   Patrin Watanatada, "Questioning and Evolving the Eco-label," *The Guardian Newspaper* March 10, 2011.

[47]   World Wildlife Fund, "Assessment of on-pack, wild capture seafood sustainability certification programmes and seafood ecolabel," December 2009 p. 112

[48]   Patrin Watanatada, "Questioning and Evolving the Eco-label," *The Guardian Newspaper* March 10, 2011.

[49]   Marine Stewardship Council, "Certified Sustainable Seafood," https://www.msc.org/get-certified/fisheries/eligible-fisheries

[50]   Harish Manwani, "Profits not always the point," *TED Talks*, January 2014

[51]   Cone Communications Research, "2015 Cone Communications Millennial CSR Study," *Cone Communications* 2015. http://www.conecomm.com/research-blog/2015-cone-communications-millennial-csr-study#download-research

[52]   Christine Gregoire, *Wikipedia*, "2004 Gubernatorial election,"
https://en.wikipedia.org/wiki/Christine_Gregoire

[53]   Tracey de Morsella, "Greenopia Rates 50 State Governors for
Environmental Responsibility," *Greenopia* 2009,
http://greeneconomypost.com/greenopia-ranks-state-governors-for-environmental-
responsibility-3709.htm

[54]   Ibid.

[55]   Governor Jay Inslee, "British Columbia, California, Oregon and
Washington join forces to combat climate change," *Washington State Governor's Office*,
Media Releases, October 2013.

[56]   Governor Jay Inslee, "Governor Inslee announces executive action to
reduce carbon pollution and promote clean energy," *Washington State Governor's Office*,
Media Releases, April 2014

[57]   Pacific Coast Collaborative, "West Coast Leaders Convene to Sign
Comprehensive Agreements at Global Clean Energy Ministerial," *News From the Region*,
June 2016

[58]   Alaska Department of Environmental Conservation, "1990-2010
Greenhouse Gas Inventory," *Climate Change in Alaska*, March 12, 2015

[59]   Amy Weinfurter, "Red states pump out more carbon pollution than
blue ones," *Grist*. June 2014. https://grist.org/climate-energy/red-states-pump-out-more-
carbon-pollution-than-blue-ones/

[60]   Wendy Koch, "Greenest US states? Some may surprise," *USA Today*
April 14, 2011.
http://content.usatoday.com/communities/greenhouse/post/2011/04/greenest-states-
surprise/1#.WBp64yQYOLU

[61]   Ibid.

[62]   Jesse Berst, "Smart Grid Leadership: The Top Ten 'Smartest' States in
2009," *Green Tech Media* March 23 2009.

[63]   Thomas Friedman, *Hot, Flat and Crowded: Why we need a green
revolution and how it can renew America*, (New York City, New York: Farrar Straus
Giroux 2008) p. 240

[64]   David Burstein. *Fast Furious: How the Millennial Generation is Shaping
our World*. (Boston, Mass. Beacon Press Books 2013), 52-53

[65]   Ibid., p. 57

[66]   Katherine Fulton, "You are the future of philanthropy," *TED Talks*,
March 2007

[67]   Bloomberg New Energy Finance, "Global Trends in Sustainable Energy
2010," *Bloomberg Finance*, 2010 p. 15

68   BNP Paribas, "BNP Paribas dedicates 15 billion euros in financing for renewable energy and reinforces its carbon risk management", *BNP Paribas the Bank for a changing world.* November 11 2015 https://group.bnpparibas/en/press-release/bnp-paribas-dedicates-e15bn-financing-renewable-energy-reinforces-carbon-risk-management-policies

69   Mount Kilimanjaro Guides, "Climate change and Kilimanjaro," *Climb Mount Kilimanjaro,* http://www.climbmountkilimanjaro.com/about-the-mountain/weather/climate-change/

70   Al Gore, "New Opportunities for Sustainable Capitalism: full text of speech," *Envirotech and Clean Energy Investor Summit,* London November 1, 2013 p. 3

71   Al Gore, "The case for optimism on climate change," *TED Talks,* February 2016

72   Al Gore, "The Turning Point", *Rolling Stone Magazine,* July 3-7, 2014 p. 86

73   Thrivenotes, "5 Great Buddha Quotes," http://www.thrivenotes.com/5-great-buddha-quotes/

74   Stephen Ambrose, "Lewis and Clark: The Journey of Corps of Discovery", *PBS Historical Series,* November 3 2016. http://www.pbs.org/lewisandclark/living/idx_9.html

75   David Hulen, "CNN on the Corrupt Bastards Club," *Alaska Dispatch News,* December 17 2007.

76   Deborah Williams, "Global Warming: The Greatest Threat", Powerpoint, *Alaska Conservation Solutions,* 2006.

77   Henry Paulson, "The Coming Climate Crash: Lessons for Climate Change in the 2008 Recession," *The New York Times,* June 21, 2014

79   Mollie Reilly, "Sarah Palin Compares Climate Change 'Hysteria' to Eugenics," *The Huffington Post,* October 27 2014.

80   Gary Funk and Brian Kennedy, "The Politics of Climate Change," *Pew Research Center,* October 4 2016. http://www.pewinternet.org/2016/10/04/the-politics-of-climate/

81   International Panel on Climate Change, "Climate Change 2014 Synthesis Report for Policymakers," *IPCC Fifth Climate Assessment*

82   Zachary Green, "In Iceland, elves aren't just Santa's little helpers," *The Rundown, PBS Newshour,* December 22, 2013.

83   Jenna Gottlieb, "Iceland's hidden elves delay road projects," *Associated Press, Juneau Empire,* December 23, 2013.

84   Ibid.

[85]   Bjork Interview, "The Colbert Report," *Comedy Central*, January 31, 2012. http://www.cc.com/video-clips/qs311n/the224-225-colbert-report-bjork

[86]   Aldo Leopold, *A Sand County Almanac* (New York: Oxford University Press 1949)

[87]   Naomi Klein. *This Changes Everything*. (New York: Simon and Schuster, 2014), 294-295.

[88]   Ibid., p. 320

[89]   Ibid. p. 361.

[90]   Joe Romm, "How to Engage and Win the Conversation about Climate and Energy," *Climate Progress*, posted November 6, 2014. https://thinkprogress.org/how-to-engage-and-win-the-conversation-about-climate-and-energy-4a06f4b826af#.q5mjo05en

[91]   Kathleen Dean Moore and Michael P. Nelson, *Moral Ground: Ethical Action for a Planet in Peril* (San Antonio: Trinity University Press 2010) p. Table of Contents

[92]   Kate Troll "Corporate interested 4, Alaska Zero. Game Over?" *Alaska Dispatch News – Guest Editorial* July 19, 2013.

[93]   Dermot Cole, "State losses on oil production tax credits spark debate on how and why," *The Alaska Dispatch News,* January 27, 2015.

[94]   Jeff Goodell, "Interview with Gus Speth: Change Everything Now," *Orion Magazine,* September/October 2008.

[95]   Ballotpedia, "Alaska Oil Tax Cut Veto Referendum, Ballot Measure 1 (August 2014)," *Ballotpedia, the Encyclopedia of American Politics*, August 19, 2014. https://ballotpedia.org/Alaska_Oil_Tax_Cuts_Veto_Referendum,_Ballot_Measure_1_(August_2014)

[96]   Suzanne Goldenberg, "ExxonMobil gave millions to climate-denying lawmakers despite pledge," *The Guardian,* July 25, 2015.

[97]   Susan Davis, "Senate affirms climate change is not a hoax," *USA Today,* January 21, 2015. http://www.usatoday.com/story/news/politics/2015/01/21/senate-climate-change-votes/22120041/

[98]   Marilyn Geewax, "New York Attorney General Investigates on Whether Exxon Mobil Lied on Climate Change," *National Public Radio Breaking News,* November 5, 2015. http://www.npr.org/sections/thetwo-way/2015/11/05/454917914/new-york-attorney-general-investigating-exxonmobil-on-climate-change

[99]   Ibid.

[100]   Douglas Fischer, "Dark Money Funds Climate Denial Effort," *Scientific American*, December 2013, https://www.scientificamerican.com/article/dark-money-funds-climate

[101]   World Resources Institute, "Understanding the IPCC Reports," *World Resources Institute – Making Big Ideas Happen,* 2015. http://www.wri.org/ipcc-infographics

[102]   Christophe McGlade and Paul Ekins, "The geographical distribution of fossil fuels unused when limiting global warming to 2 degrees Celsius," *Nature Issue 517* January 8, 2015

[103]   Ibid.

[104]   Common Dreams, "The Mother of All Risks: Insurance Giants Call on G20 to Stop Bankrolling Fossil Fuels," *Breaking News & Views for the Progressive Community,* September 1, 2016.

[105]   Naomi Klein. *This Changes Everything.* (New York: Simon and Schuster, 2014), 353-355

[106]   350.org, "Fossil Free Campaign", *350.org* http://gofossilfree.org/commitments/

[107]   Damian Carrington, "World's biggest sovereign wealth fund dumps dozens of coal companies," *The Guardian*, February 5, 2015. https://www.theguardian.com/environment/2015/feb/05/worlds-biggest-sovereign-wealth-fund-dumps-dozens-of-coal-companies

[108]   Naomi Klein. *This Changes Everything.* (New York: Simon and Schuster, 2014), 355

[109]   Common Dreams, "Global Disinvestment Day Challenges Fossil Fuel Industry on February 13-14," *Breaking News & Views for the Progressive Community,* February 13, 2015. http://www.commondreams.org/newswire/2015/02/13/global-divestment-day-challenges-fossil-fuel-industry-february-13-14

[110]   Eric Lipton, "Hard-Nosed Advice from Veteran Lobbyist: 'Win Ugly or Lose Pretty', Richard Berman Energy Industry Talk Secretly Taped," *New York Times*, October 30, 2014.

[111]   Environmental Policy Alliance, "Divest From Fossil Fuels? Not if you Like Your Life," *Big Green Radicals: Updates*, February 10 2015. https://www.biggreenradicals.com/divest-in-fossil-fuels-not-if-you-like-your-life/

[112]   Naomi Klein. *This Changes Everything.* (New York: Simon and Schuster, 2014), 296

[113]   Business Page, "Greens in pinstriped suits," *Economist Magazine* May 21, 2016. p. 57

114   Naomi Klein. *This Changes Everything*. (New York: Simon and Schuster, 2014), 343

115   Peter Russell, *Waking Up in Time*. (Sab Rafael, CA: Origin Press, 2009) p. 36

116   Alaska Department of Fish and Game, "Status and Trends of Humpback Whales," *Species Reports*, http://www.adfg.alaska.gov/index.cfm?adfg=humpbackwhale.main

117   James Gustave Speth, *The Bridge at the Edge of the World: Capitalism, The Environment, and Crossing from Crisis to Sustainability*,(New Haven and London: Yale University Press, 2008) p. 236

118   David Burstein. *Fast Furious: How the Millennial Generation is Shaping our World*. (Boston, Mass. Beacon Press Books 2013), p. 94.

119   Ibid., p. 95

120   Harvard Institute of Politics, "Spring 2016 Poll," *Youth Polls*, April 2016 http://iop.harvard.edu/youth-poll/harvard-iop-spring-2016-poll

121   Al Gore, *The Future: Six Drivers of Global Change*, (New York: Random House, 2013) p. 5-8.

122   Ibid.

123   Ibid. p. 9.

124   Oxfam International, "Wealth: Having it All and Wanting More," *Research Reports* January 19, 2015. https://www.oxfam.org/en/research/wealth-having-it-all-and-wanting-more

125   Bloomberg Businessweek, "Capitalism Seen in Crisis by Investors Citing Inequalities," *Bloomberg Poll*, January 25, 2012.

126   Rana Foroohar, "American Capitalism's Great Crisis," *Time Magazine*, May 12, 2016.

127   Ibid.

128   Al Gore, *The Future: Six Drivers of Global Change*, (New York: Random House, 2013) p. 365

129   Nic Marks, "The Happy Planet Index," *TED Talks*, August 2010 https://www.ted.com/talks?sort=newest&q=nic+marks

130   Laura Musikanski, "The UN Embraces the Economics of Happiness," *Yes Magazine*, April 12, 2012.

131   UN Member States of Sustainable Development Goals, *World Happiness Report 2013*, Summary

132   Ibid. Chapter 5, Values and Happiness.

133   James Gustav Speth, *America the Possible: Manifesto for a New Economy*, (New Haven and London: Yale University Press, 2012) p. 85

134   Daniel Kahneman and Angus Deaton, "High income improves evaluation of life but not emotional well-being," *Proceedings of the National Academy of Sciences* Volume 107, No. 38 August 4 2010.

135   Yereth Rosen, "Acidification take toll on Beaufort Sea; threats loom in Chukchi and Bering," *Alaska Dispatch News*, June 15, 2015.

136   Craig Welch, "Sea Change: Vital Part of Web Dissolving," *The Seattle Times*, April 30th 2014 http://apps.seattletimes.com/reports/sea-change/2014/apr/30/pteropod-shells-dissolving/

137   Jeremy Mathis and Steve Colt, "A wake up call in Alaska about ocean acidification and coastal communities," *Alaska Dispatch News*, July 29, 2014. https://www.adn.com/commentary/article/wake-call-alaska-about-ocean-acidification-and-coastal-communities/2014/07/29/

138   Barak Obama, "Full Remarks of the President's Speech at GLACIER Conference," *Alaska Dispatch News*, August 31, 2105.

139   National Park Service, "Grand Canyon Park Statistics", *Grand Canyon National Park Service*. https://www.nps.gov/grca/learn/management/statistics.htm

140   Connie Barlow, "The Great Unconformity of the Grand Canyon: Sacred Site of the Epic of Evolution," August 2009. http://thegreatstory.org/great-unconformity.html

141   Bill Plotkin, *Nature and the Human Soul* (Novato, California: New World Library 2008) p. 35-40

142   Anthropocene, *Wikipedia* "Anthropocene era". https://en.wikipedia.org/wiki/Anthropocene

143   Al Gore, "The Turning Point", *Rolling Stone Magazine*, July 3-7, 2014 p. 86

144   Jeff Goodell, "Will the Paris Climate Deal Save the World?" *Rolling Stone Magazine,* January 13, 2016. p.33

145   Al Gore, *The Future: Six Drivers of Global Change*, (New York: Random House, 2013) p. 361

146   James Gustave Speth, *The Bridge at the Edge of the World: Capitalism, The Environment, and Crossing from Crisis to Sustainability*, (New Haven and London: Yale University Press, 2008) p. 202-206.

147   Ibid. p. 208-209.

148   Al Gore, "The case for optimism on climate change," *TED Talks,* February 2016.

About the Author

Kate Troll is an op-ed columnist, wilderness adventurer, and speaker on conservation and climate issues. She moved to Alaska in 1977, seeking wilderness and a career in natural resource management.

As Executive Director of the Alaska Conservation Voters, Kate helped draft the creation of the Alaska Renewable Energy Fund and lobbied for the Sustainable Energy Act, a comprehensive roadmap to generate 50% of Alaska's electrical energy from renewable sources by 2025. She served as Executive Director for United Fishermen of Alaska (nation's largest fishing organization). She also worked as a fisheries development specialist and policy analyst for the State of Alaska. Internationally, Kate was Regional Fisheries Director (North and South America) for the Marine Stewardship Council, a global eco-label program.

She was elected to public office twice, serving on the Juneau-Douglas Borough and Ketchikan Borough Assembly. Kate was appointed by Governor Palin to serve on the Alaska Climate Mitigation Advisory Board, and was the only Alaskan invited to participate in Governor Schwarzenegger's 2008 Global Climate Summit.

From 2010-2016, she wrote for the *Juneau Empire* and is currently a columnist for the *Alaska Dispatch News*. She is writing her third screenplay, *Hot Ice, Cold Lies*, drawing upon her activist career and wilderness adventures.

Kate has skied the Great Gorge of Ruth Glacier, run the Alsek-Tatshenshini Rivers, climbed Mount Denali and Mount Kilimanjaro, rafted through the Grand Canyon in winter, and has kayaked Alaska's outside coast on several occasions. Her explorations provide inspiration for stories that reflect her deep love of place and the wisdom to be found there.